QGIS and Applications in Water and Risks

QGIS in Remote Sensing Set

coordinated by
André Mariotti

Volume 4

QGIS and Applications in Water and Risks

Edited by

Nicolas Baghdadi
Clément Mallet
Mehrez Zribi

iSTE

WILEY

First published 2018 in Great Britain and the United States by ISTE Ltd and John Wiley & Sons, Inc.

ISTE Ltd
27-37 St George's Road
London SW19 4EU
UK

www.iste.co.uk

John Wiley & Sons, Inc.
111 River Street
Hoboken, NJ 07030
USA

www.wiley.com

Library of Congress Control Number: 2017962518

British Library Cataloguing-in-Publication Data
A CIP record for this book is available from the British Library
ISBN 978-1-78630-271-7

MIX
Paper from responsible sources
FSC® C013604

Contents

Introduction

In the context of global changes (climatic and anthropogenic), the understanding and the quantification of the dynamics of water resources and risks of all types are essential for the managers of the territories in order to propose the most efficient solutions for planning and adaptation to changes. Geographic Information Systems (GIS) are invaluable tools in this domain. This is due to the sharply increasing use of remote sensing data, measurement networks and now products available to end users (several Copernicus services covering these issues are thus now mature).

This volume is devoted to the presentation and implementation under QGIS and its libraries of case studies dealing with water and risk issues. Chapters 1 and 2 discuss the application to bathymetry mapping. Chapters 3–5 are devoted to hydrological problems, both from a surface (reservoirs) and a linear (networks) point of view. Chapters 6–8 are dedicated to the presentation of three types of risks arising from very different hazards (drought, crop pests, erosion).

This volume, supported by scientists internationally renowned in their fields, aims to update readers' knowledge and to describe the main research and development issues for forthcoming years. It is intended for research teams in geomatics, second-level (engineering schools, master's degrees) and postgraduate studies (PhD students), and engineers involved in the management of water and territory resources. In addition to the texts of the proposed chapters, the readers will have access to the data and tools, as well as all screenshots related to the QGIS application in order to smoothly carry out each step of each application.

A supplement to the chapters, including datasets and screenshots illustrating the practical application of the chapters, is available at the following address:

Using Internet Explorer: ftp://193.49.41.230
Using a FileZilla client: 193.49.41.230

Username: **vol4_en**
Password: **334@Volne**

We would like to thank everyone who contributed to the development of this book; first of all, the scientists, authors of the chapters of course, but also the experts of the scientific committee for their inspection of the chapters and the corrections made. This project was carried out thanks to the support of IRSTEA (French Research Institute for Science and Technology for Environment and Agriculture), CNRS (French National Center for Scientific Research), IGN (National Institute for Geographic and Forest Information) and CNES (French National Center for Space Studies).

We are very grateful to Airbus Defense and Space and Equipex Geosud for providing us with SPOT 5/6/7 images. Please note that these images may only be used in a research and training framework and any commercial activity based on the data provided is strictly prohibited.

Our thanks also go to our families for their support in realizing this project and to André Mariotti (Professor Emeritus, Pierre and Marie Curie University) and Pierrick Givone (President, IRSTEA) for their encouragement.

Nicolas BAGHDADI

Clément MALLET

Mehrez ZRIBI

1

Monitoring Coastal Bathymetry Using Multispectral Satellite Images at High Spatial Resolution

1.1. Definition, context and objective

In this chapter, the instantaneous water depth (*H*), which assumes a negative value, is defined as the vertical distance from the water surface to the bottom at a given time. In coastal shallow waters associated with tidal environments, the variations of *H* depend first on the phase of the tide. The water height due to the tide (*T*), which usually assumes a positive value, can be calculated from a local tidal datum, which provides standard elevations with respect to a reference level (T_0; for instance, T_0 is defined as the lowest astronomical tide for the France Hydrographic Service, SHOM).

The bathymetry (*Z*) is the vertical distance from the water surface to the bottom, corrected from the tide level. For more information, the reader can refer to the publication of the International Hydrographic Organization on the standards for hydrographic surveys[1]. Following the previous definitions, the relationship between *Z*, *H* and *T* is then given by:

$$Z = H + T \qquad [1.1]$$

Chapter written by Bertrand LUBAC.

1 https://www.iho.int/iho_pubs/standard/S-44_5E.pdf.

When $|H| > |T|$, the value of Z is negative meaning that the bottom elevation is lower than the reference elevation of the tidal datum (Figure 1.1, case 1). In the other cases, Z assumes a positive value that means the bottom elevation is higher than the reference elevation (Figure 1.1, case 2).

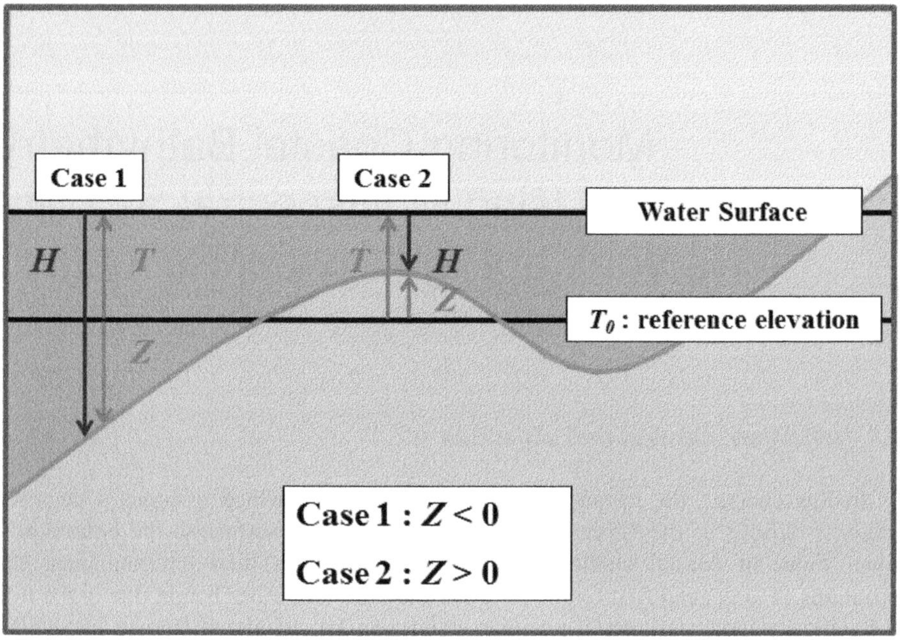

Figure 1.1. *Schematic description of the main parameters used to characterize the water height in coastal shallow waters: instantaneous water depth (H), tidal height at a given time (T) and bathymetry (Z). For a color version of the figure, see www.iste.co.uk/baghdadi/qgis4.zip*

Bathymetry is a key environmental parameter for studying and modeling the hydrodynamics, the sediment transport and the long-term morphological evolution of coastal systems as well as for predicting their responses to hydroatmospheric maritime hazards and assessing their vulnerabilities to environmental risks. The accurate and regular measurement of bathymetry is therefore a major scientific issue with strong socioeconomic implications.

Different systems of observation can be used to measure bathymetry. Active sensors (single or multibeam echo sounder and bathymetric Lidar[2]), deployed on

2 Light detection and ranging.

seaborne or airborne platforms, allow for the implementation of robust methodologies that provide high-quality products. Passive optical sensors (multispectral radiometers and hyperspectral cameras), mainly mounted onboard air- or spacecraft, require the use of recent methodologies which are usually still in development. Bathymetry associated with these systems of observation generally presents a higher uncertainty than the products retrieved from active sensors. However, multi- and hyper-spectral missions provide data at lower cost with a higher temporal acquisition frequency as well as a larger sample surface.

These different systems of observation provide highly complementary data, which is useful to address operational and research issues. The current development of methods of multisource bathymetry reconstruction promotes the joint exploitation of observations of different origins and qualities. This chapter, which describes a simple and robust method for the inversion of bathymetry, is written from this perspective.

Following this, the objective of this chapter is to present a simple method which allows for the mapping of the bathymetry in the inlet of the Arcachon lagoon as well as the characterization of the temporal evolution of the sediment stock, using multispectral satellite images. The method was validated by scientific studies, while the derived products are commonly used by decision makers. The method is selected partly on pedagogical criteria and is fully adapted to non-specialized readers. In the following sections, the method is applied for the processing of two Sentinel-2/MultiSpectral Instrument (MSI) images, which were acquired with a one-year interval.

1.2. Description of the methodology

Two approaches can be used to measure the bathymetry from multispectral satellite images, namely, the empirical and the semianalytical approach [CAP 14]. The empirical approach is based on statistical models, which are established using field bathymetry data and top-of-atmosphere (ρ_{TOA}) or surface (ρ_w) reflectances derived from satellite measurements located at the same geographical location. The semianalytical model is based on physical models, which allow for the calculation of H as a function of ρ_w. However, the calibration of some parameters of the physical model may require an optimization. This procedure is usually based on a training data set derived from field measurements.

The advantages of the empirical approach are the simplicity of the implementation of the algorithm and the possibility of directly using ρ_{TOA}. This last point is particularly important for aquatic color applications in coastal waters. The calculation of ρ_w requires the use of specific and complex corrections in order to remove the atmospheric contributions to ρ_{TOA}. The main disadvantage is the necessity of calibrating the statistical model for each image from a field dataset, which must be statistically representative of the study area. Furthermore, empirical approaches usually assume that the optical properties of the substrate and the water column are homogeneous.

For the case study presented in this chapter, the empirical approach was selected for the following reasons:

– the advantages mentioned previously;

– the availability of field bathymetry data to calibrate the statistical model used for the bathymetry inversion;

– the quasi-spatial homogeneity of the optical properties over the study site.

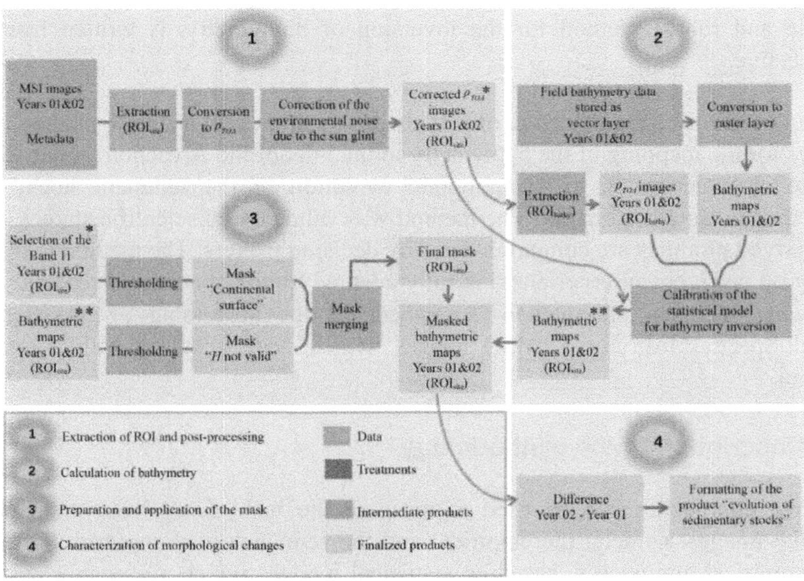

Figure 1.2. *Schematic flowchart of image processing under QGIS for bathymetry mapping. For a color version of the figure, see www.iste.co.uk/baghdadi/qgis4.zip*

The mapping of the bathymetry is based on a succession of different treatments illustrated in Figure 1.2. To facilitate the reading, the treatments are grouped into four main steps, ranging from the correction of the image noise due to environmental effects (sun glint) to the characterization of the morphological evolution of the main sedimentary structures:

1) selection and downloading of the images, extraction of the region of interest (ROI) and treatment of the environmental effects;

2) conversion of bathymetry data stored in vector format to raster format and calibration of the statistical model;

3) preparation and integration of the masks associated with the emerged continental surfaces and the aquatic surfaces for which the water depth is outside the validity range;

4) characterization of the morphological evolution of the main sedimentary structures.

1.2.1. *Step 1: selection and preprocessing of MSI images*

The first step is to select and download the MSI images as well as to extract the ROI and correct ρ_{TOA} from the environmental noise. The criteria used to select the MSI images are (1) suitable atmospheric conditions, namely, low percentage of cloudy pixels and low aerosol optical depth and (2) homogeneous seawater optical properties, which means no plume, for instance, due to particle resuspension, and no phytoplankton bloom. MSI images can be downloaded from different public platforms, which provide free and open access to the Sentinel-2 user products. Here, we propose using the Copernicus Open Access Hub[3] to download the Level-1C[4] product. This product provides orthorectified ρ_{TOA} with subpixel multispectral registration. ρ_{TOA}, with values often ranging from 0 to 1, has been converted into integer values (IV; coded in 12 bits) by applying a fixed coefficient (1,000 by default). The Level-1C image corresponds to a set of tiles. A tile is a 100 km by 100 km squared orthoimage in UTM/WGS84 projection. There is one tile per spectral band. One Level-1C image including all bands is approximately 500 MB in file size. When the study area is smaller than the tile dimension, it is recommended to extract the ROI of the site (ROI_{site}) in order to decrease the file size.

> **QGIS functions for ROI_{site} extraction and conversion from IV to ρ_{TOA}:**
> - ROI_{site} Extraction: *Raster > Extraction > Clipper...*
> - Conversion from IV to ρ_{TOA} : *Raster > Raster Calculator...*

High spatial resolution satellite images (O ~10 m) associated with littoral aquatic systems generally have strong environmental noise, which has to be corrected before any analysis. This natural noise is due to the impact of different environmental factors, which generate an interpixel variability of ρ_{TOA} independent of the change in depth, seawater and substrate optical properties. One of the main factors limiting

3 https://scihub.copernicus.eu/.

4 https://sentinels.copernicus.eu/web/sentinel/technical-guides/sentinel-2-msi/level-1c-processing.

the accuracy of remotely sensed data from a littoral environment is sun glint, the specular reflection of light from seawater surfaces (Figure 1.3).

Different approaches have been developed to correct ρ_{TOA} from sun glint. A comprehensive bibliographic synthesis can be found in [BRU 15]. In this chapter, the sun glint correction method of Hedley *et al.* [HED 05] was selected. The sun glint correction procedure is based on the following steps:

1) extraction of an ROI associated with optically homogeneous and deep[5] waters ($ROI_{sunglint}$);

2) determination of the minimum value of $\rho_{TOA}(NIR)$ (*MIN*) and the slope value (b) of the linear regression between $\rho_{TOA}(VIS)$ (dependent variable) and $\rho_{TOA}(NIR)$ (explanatory variable). NIR and VIS are associated with the spectral bands located in the near-infrared and visible wavelengths, respectively;

3) correction of $\rho_{TOA}(VIS)$ over the whole ROI by applying the following formula:

$$\rho_{TOA}^{cor}(VIS) = \rho_{TOA}(VIS) + b\big(MIN - \rho_{TOA}(NIR)\big) \quad [1.2]$$

where $\rho_{TOA}^{cor}(VIS)$ is the sun glint-corrected reflectance at the visible spectral bands. In the following sections, we will use ρ_{TOA}^{cor} for the retrieval of Z. Thus, the suffix "*cor*" will no longer be explicitly mentioned in order to facilitate the reading of the text.

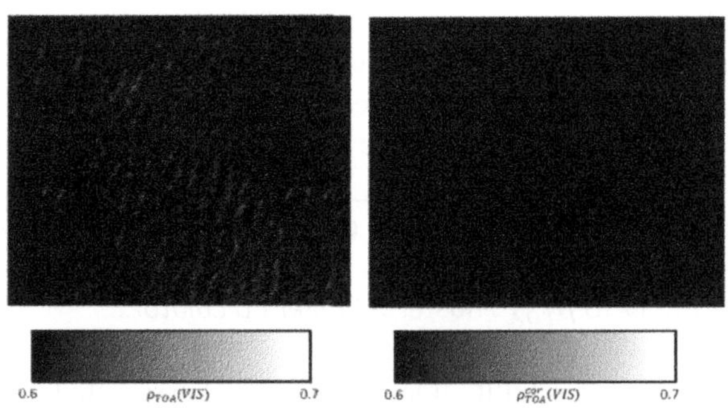

Figure 1.3. *On the left, example of an MSI image acquired in the band 3 (560 nm), which shows the sun glint impact on $\rho_{TOA}(VIS)$. On the right, the sun glint-corrected image ($\rho_{TOA}^{cor}(VIS)$) using the approach of Hedley et al. [HED 05]*

5 Optically deep waters define the marine pixels for which the radiative signal emitted by the aquatic surface is not influenced by the bottom.

QGIS functions for the sun glint correction:
- $ROI_{sunglint}$ Extraction: *Raster > Extraction > Clipper*
- Calculation of the linear regression slope *b*: *Processing > Toolbox > GRASS GIS 7 commands > r.regression.line*
- Sun glint correction: *Raster > Raster Calculator...*

1.2.2. *Step 2: calibration of the bathymetry inversion model*

The second step is the calibration of the empirical model used to retrieve the bathymetry (Z) using multispectral images. The empirical model is based on the analytical formulation of Maritorena *et al.* [MAR 94], which establishes the relationship between the instantaneous water depth (H) and the water-leaving reflectance (ρ_w):

$$H = -\frac{1}{2K}\left[ln\left(\rho^{bottom} - \rho_w^{deep}\right) - ln\left(\rho_w - \rho_w^{deep}\right)\right] \quad [1.3]$$

where K (m^{-1}) is the vertically-averaged effective or operational attenuation coefficient, ρ^{bottom} is the bottom reflectance, and ρ_w^{deep} is the reflectance over optically deep waters. For optically homogeneous areas, the values of K, ρ_w^{deep}, and ρ^{bottom} can be considered as constants. In this case, equation [1.3] can be simplified in order to obtain an empirical model given by the following formulation:

$$Z = \alpha + \beta \times ln\left(\rho_{TOA}(G) - \rho_{TOA}^{deep}(G)\right) \quad [1.4]$$

where G is a spectral band located in the green wavelengths; and α and β are constants that must be calculated for each image using a linear regression between field bathymetry data and $\rho_{TOA}(G)$. The selection of the G-band is based on the fact that K is minimum in these wavelengths allowing the calculation of Z for larger water depths. ρ_w is replaced by ρ_{TOA}. This simplification is based on the fact that we assume a linear relationship between the two parameters. Similarly, H is replaced by Z (equation [1.1]). Here, we also assume that the water height due to the tide (T) is quasi-constant over the whole area for a given image. This last assumption is not completely true. At a given time, T varies in the Arcachon inlet from a few centimeters to a few tens of centimeters according to the meteorological and tide conditions. This assumption generates an uncertainty of a few percent on the bathymetry retrieval, which is assessed as reasonable with regard to the other sources of error [CAP 14].

The approach taken to calibrate the empirical model is as follows:

1) conversion of the field bathymetry data distributed as vector images into raster images;

2) determination of the $\rho_{TOA}^{deep}(G)$ value using the subimage of $\rho_{TOA}(G)$ associated with the $ROI_{sunglint}$;

3) from the ROI_{site}, extraction of the ROI associated with the geographical limits of the raster image of the field bathymetry data (ROI_{bathy});

4) calculation of $ln\left(\rho_{TOA}(G) - \rho_{TOA}^{deep}(G)\right)$ for the sub-image of $\rho_{TOA}(G)$ associated with the ROI_{bathy};

5) determination of the α and β constants (equation [1.4]) using a linear regression;

6) calculation of the bathymetry using the empirical model over the whole study site.

It is important to note that we do not propose to the reader to quantify the error performed on the bathymetry retrieval. The reason is that we do not have additional data to perform the validation exercise for this data set.

QGIS functions for the calibration of the empirical model:
- Format conversion of the field bathymetry data: *Raster > Conversion > Rasterize*
- ROI_{bathy} extraction: *Raster > Extraction > Clipper...*
- Calculation of $ln\left(\rho_{TOA}(G) - \rho_{TOA}^{deep}(G)\right)$: *Raster > Raster Calculator...*
- Calculation of the α and β constants: *Processing > Toolbox > GRASS GIS 7 commands > r.regression.line*
- Calculation of the bathymetry : *Raster > Raster Calculator...*

1.2.3. *Step 3: preparation and application of the masks*

The third step is dedicated to the preparation of two masks, which will then be applied on the bathymetry maps. The objective is to mask pixels that are associated with emerged continental surfaces ("continental surface") and aquatic surfaces for which the water depth is outside the validity range ("*H* not valid"). The masked pixels will be coded as "*no data*" values.

The "continental surface" mask is based on the use of band 11 of MSI, which is located in the shortwave infrared (SWIR) at approximately 1,610 nm. For littoral environments, the absolute frequency histogram of $\rho_{TOA}(SWIR)$ is generally characterized by a bimodal distribution. The first (second) mode, which shows low (high) values of $\rho_{TOA}(SWIR)$, can be associated with aquatic surfaces (continental surfaces) due to the fact that the radiative signal measured for marine pixels depends solely on the atmospheric component, through the molecular and aerosol diffusion process of the incident solar radiation flux. The surface component, associated here with the marine signal, is assumed to be zero due to a very high absorption of light by pure sea water. Thus, from an analysis of the histogram, a threshold value can be determined that makes it possible to differentiate the aquatic pixels from the non-aquatic pixels. It should be noted that the spatial resolution of the SWIR band is 20 m, whereas that of the VIS and NIR bands is 10 m. The SWIR image must therefore be resampled to 10 m before any processing.

The "*H* not valid" mask is associated with the validity domain of the bathymetry product and is based on the analysis of the error propagation in the inversion model, carried out by Capo *et al.* [CAP 14]. In general, the greatest error is observed for very small water depths, on average less than 1 m, and for intermediate depths, on average higher than 4 m. For shallow waters, the main factor, which explains the error, is the high resuspension of the sediment. This makes the assumption of spatial homogeneity of the optical properties of the water column inappropriate. For intermediate waters, the main factor is the error performed on the estimation of the parameters of the inversion model that propagates nonlinearly in the model.

QGIS functions for the mask processing:
- Resampling of the SWIR image on a finer grid: *Processing > Toolbox > GRASS GIS 7 Commands > Raster > r.resamp.interp*
- Tresholding: *Raster > Raster Calculator...*
- Definition of the no data values: *Raster > Projections > Warp (Reproject)...*

1.2.4. *Step 4: characterization of the morphological evolution of the main sedimentary structures*

One of the simplest methods used to characterize and quantify variations in submarine topography is based on the mapping of bathymetry for two distinct dates and then calculating their difference. This makes it possible to estimate the evolution of the sedimentary stock in terms of erosion or accretion. This approach requires a suitable formatting of results in order to highlight the variation of bathymetry.

QGIS functions for the study of the seabed morphology:
- Quantification of the sedimentary stock evolution: *Raster > Raster Calculator...*
- Formatting results: *Layer > Properties > Style*

1.3. Practical application

This section presents a practical application for the mapping of bathymetry and the characterization of the morphological evolution of the underwater sedimentary structures of the Arcachon inlet (France) using a pair of Sentinel-2/MSI images acquired at almost 1 year apart (August 28, 2015 and October 21, 2016).

1.3.1. *Software and data*

1.3.1.1. *Software required*

The matrix and vector image processing that will be performed in the following sections uses the basic functionality of the QGIS software (version 2.18). This choice allows anyone to apply processing without installing extensions. Of course, some steps of the method can be carried out with more efficient tools, in particular the Orfeo ToolBox image processing library (OTB), developed by CNES (*Centre National d'Etudes Spatiales*).

1.3.1.2. *Input data*

This exercise is based on free and open access data, which can be downloaded from different websites. These data are composed of (1) Sentinel-2/MSI (Level 1C) TOA reflectance images generated by the Sentinel-2 ground segment and (2) bathymetry measurements acquired in the framework of the MORITO project funded by the TOSCA program (CNES) during field surveys.

Downloading of Sentinel-2/MSI images

For the proposed exercise, the first step is the selection of a pair of suitable images for which different environmental criteria must be considered (Table 1.1): (1) low cloud cover; (2) high tide (T), which means the immersion of the main sedimentary structures; (3) low vertical attenuation (K) to ensure that the radiometric signal of the substrate is well resolved; (4) spatially homogeneous water column properties; and (5) low surface roughness to limit the impact of sun glint. The tile associated with the Arcachon basin has the identifier T30TXQ. Since the launch of Sentinel-2 in June 2015, only five images show favorable atmospheric conditions. General information associated with these images is given in Table 1.1. The analysis of the different criteria allows the selection of the images acquired on August 28, 2015 and October 21, 2016.

Date (dd/mm/yy)	Hour (UTM)	*T* (m)	qK (0–3)	HOP (0–3)	SR (0–3)	Quality (0–3)	Selection
28/08/2015	11:06:56	2.14	0	0	2	1	Yes
21/10/2016	11:03:18	2.44	1	0	0	0	Yes
30/11/2016	11:04:18	0.56	1	1	0	1	No
18/02/2017	11:01:25	2.87	3	1	0	2	No
10/03/2017	10:59:51	NA	3	3	2	3	No

Table 1.1. *General information associated with the preselected Sentinel-2/MSI images (T30TXQ tile) allowing their identification and selection for the bathymetry mapping exercise. Date (dd/mm/yy): date of acquisition of the image; Time (UTM): time of acquisition of the image; T (m): tidal water height; qK, HOP, SR: image quality indicators related to the penetration of light into the water column, the spatial homogeneity of the optical properties of the water column, and the sea surface roughness, respectively (values range from 0 to 3, 0 for good condition and 3 for bad condition); Quality: image quality based on qK, HOP and SR variables (values range from 0 to 3, 0 for good quality and 3 for poor quality); Selection: final decision for the selection of the images on the basis of the T and Quality variables*

The selected Sentinel-2/MSI images (Figure 1.4) can be downloaded from the Copernicus open access hub:

https://scihub.copernicus.eu/dhus/#/home

– **Sentinel-2/MSI – 28/08/2015**

File name: S2A_OPER_MSI_L1C_TL_EPA__20161006T143833_A000949_T30TXQ_N02.04

– **Sentinel-2/MSI – 21/10/2016**

File name: S2A_OPER_MSI_L1C_TL_SGS__20161021T162718_A006955_T30TXQ_N02.04

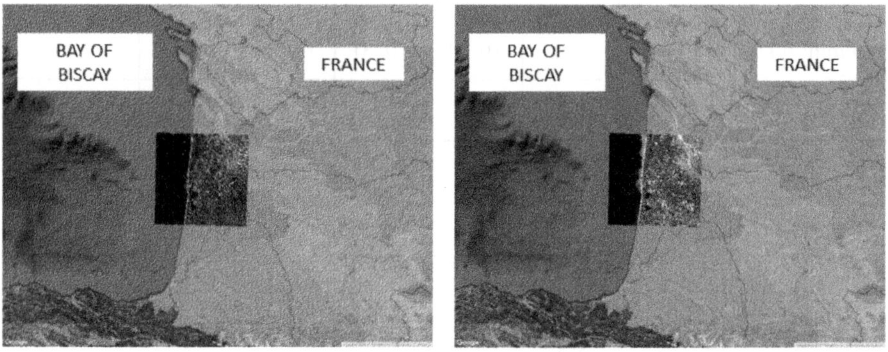

Figure 1.4. *Selected sentinel-2/MSI images acquired on 28/08/2015 (on the left) and 21/10/2016 (on the right) projected on Google Maps and BD Carthage® - © IGN-MEDD map background. For a color version of the figure, see www.iste.co.uk/baghdadi/qgis4.zip*

Each distributed Level-1C product is composed of 13 tiles associated with the spectral bands of the MSI sensor. For the exercise, only the spectral bands 3, 8 and 11 (G, NIR and SWIR channels, respectively) will be used. The instrumental characteristics of these bands are given in Table 1.2. To improve readability in the following sections, the associated images are renamed:

– MSI_2015_B03.jp2, MSI_2015_B08.jp2, MSI_2015_B11.jp2;

– MSI_2016_B03.jp2, MSI_2016_B08.jp2, MSI_2016_B11.jp2.

When processing is performed on these six images associated with two dates and three spectral bands, the generic name MSI_YEAR_BN.jp2 (BN for band number) will be used.

Band Number	Nominal center (nm)	Width (nm)	Spatial resolution (m)	SNR
3	560	35	10	154
8	842	115	10	172
11	1,610	90	20	100

Table 1.2. *Information on spectral characteristics, signal-to-noise ratio (SNR) and spatial resolution associated with the Sentinel-2/MSI 3, 8 and 11 bands (G, NIR and SWIR channels, respectively)*

Downloading of field bathymetry data acquired on August 2015 and 2016

File name: BATHY_2015.shp and BATHY_2016.shp

The vector layers are available in the additional material of this book.

1.3.2. *Step 1: extraction of the region of interest and preprocessing*

1.3.2.1. *Extraction of ROI_{site}*

The study area is smaller than the tile surface. For each tile, we recommend extracting a subimage associated with the ROI of the study site (ROI_{site}). For the following, we will use the generic name MSI_YEAR_BN.jp2 for the MSI images, where YEAR takes the values 2015 or 2016, and BN takes the values 03, 08 or 11.

Step	QGIS processing
1. ROI_{site} Extraction	In **QGIS**: • Open the image MSI_YEAR_BN.jp2 In the **Menu Bar**: • Select **Raster > Extraction > Clipper** In the **Clipper Dialog Box:** • Input file: - Select .../MSI_YEAR_BN.jp2 • Output file: - Select .../MSI_YEAR_BN_SITE.TIF • **Clipping Mode**: - Select the option **Extent** • Click on the button **Edit** : - Draw a rectangle delimiting the Arcachon inlet and adjacent shorelines *Example: top left pixel (E = 624,774 m, N = 4,949,974 m); bottom right pixel (E = 648,655 m, N = 4,926,274 m).* • Click on the **OK** button *Note: do not click on the* ***Close*** *button until you have extracted all the sub-images.*

1.3.2.2. *Preprocessing of the subimages*

ρ_TOA are converted and distributed into IV. To convert them into reflectance values, IV have to be divided by 1,000. Then, reflectance values have to be corrected for the contamination of sun glint as well as wave breaking and watercraft (Figure 1.5). The method of Hedley *et al.* [HED 15] can be used to correct these artifacts.

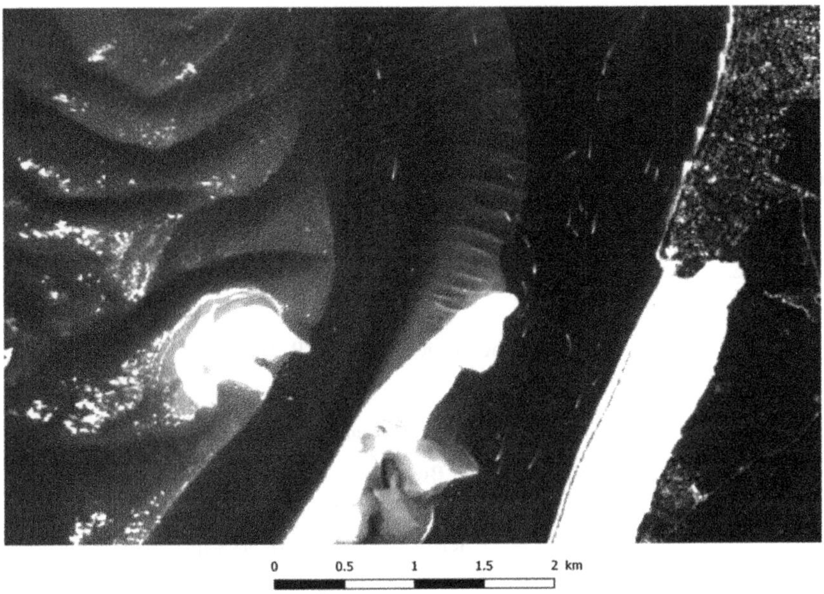

Figure 1.5. *Zoom on the Arcachon inlet (coordinates of the center point: E = 640, 136 m; N = 4,939,527 m) of the Sentinel-2/MSI image acquired on 28/08/2015, which allows the illustration of the marine pixel contamination due to sun glint as well as wave breaking and watercraft*

Step	QGIS processing
1. Conversion of IV to reflectance values	In **QGIS**: • Check the presence of the following raster files: - MSI_YEAR_BN_SITE.TIF In the **Menu Bar**: • Select **Raster > Raster Calculator...** Properly recharge the current layer extent in order to avoid unexpected resampling, which can cause problems later.

<table>
<tr>
<td></td>
<td>In the Raster Calculator Dialog Box:
• Raster calculator expression:
"MSI_YEAR_BN_SITE@1"/1000
• Output layer:
- Select .../MSI_YEAR_BN_RHO.TIF</td>
</tr>
<tr>
<td>2. Correction of the sun glint</td>
<td>In QGIS:
• Check the presence of the following raster files:
- MSI_YEAR_B03_RHO.TIF
- MSI_YEAR_B08_RHO.TIF
In the Menu Bar:
• Select Raster > Extraction > Clipper</td>
</tr>
<tr>
<td>2.1. Extraction of $ROI_{sunglint}$</td>
<td>In the Clipper Dialog Box:
• Input file:
- Select the raster layers successively
• Output file:
- Select .../MSI_YEAR_BN_DEEP.TIF
• Clipping Mode:
- Select the option Extent
• Click on the button Edit :
- Draw a rectangle delimiting an optically deep water area located in the Arcachon inlet
Example: top left pixel (E = 639,704 m, N = 4,943,394 m); bottom right pixel (E = 640,114 m, N = 4,941,434 m) (red area in Figure 1.6).
• Click on the OK button</td>
</tr>
</table>

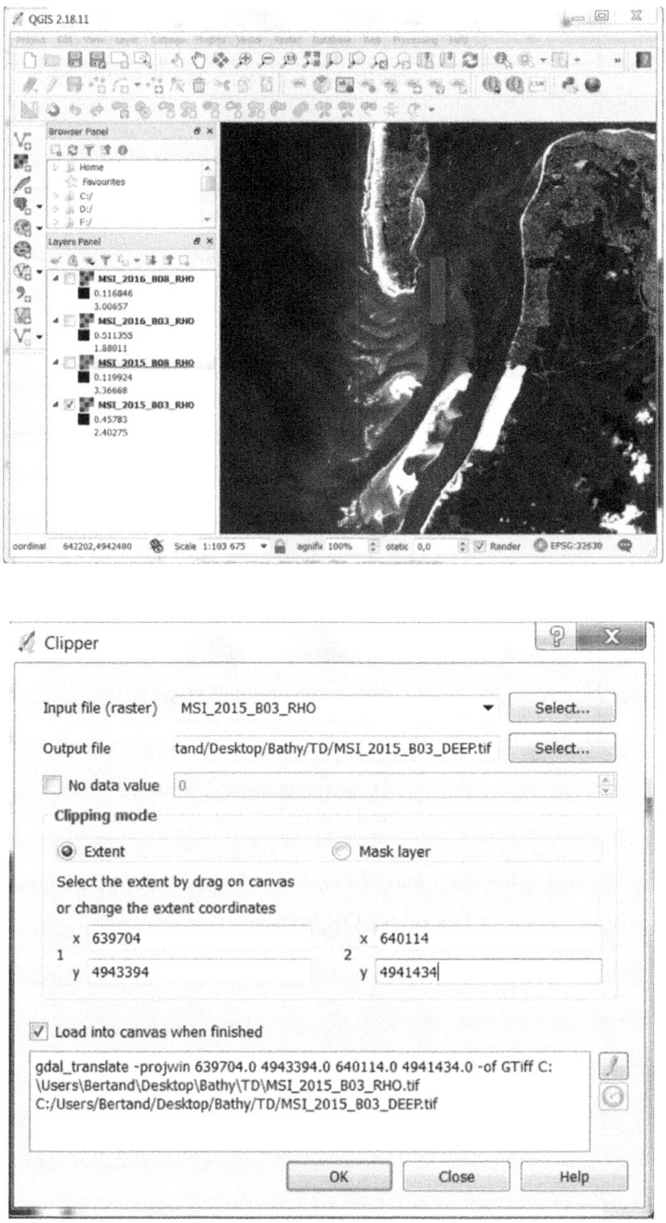

Figure 1.6. *Illustration of the QGIS processing used to extract the ROI*$_{sunglin}$*tm. For a color version of the figure, see www.iste.co.uk/baghdadi/qgis4.zip*

2.2. Calculation of *MIN* and *b* parameters	Determination of ***MIN*** values for 2015 and 2016 (Figure 1.7) • In the **Layers Panel**, for layers named MSI_YEAR_B08_DEEP.TIF, **read the minimum value of** $\rho_{TOA}(NIR)$ *Example of MIN values: 0.1600 for 2015 and 0.1320 for 2016 (for* ρ_{TOA}*, the fifth decimal is not significant. Values can be rounded to the fourth decimal).*

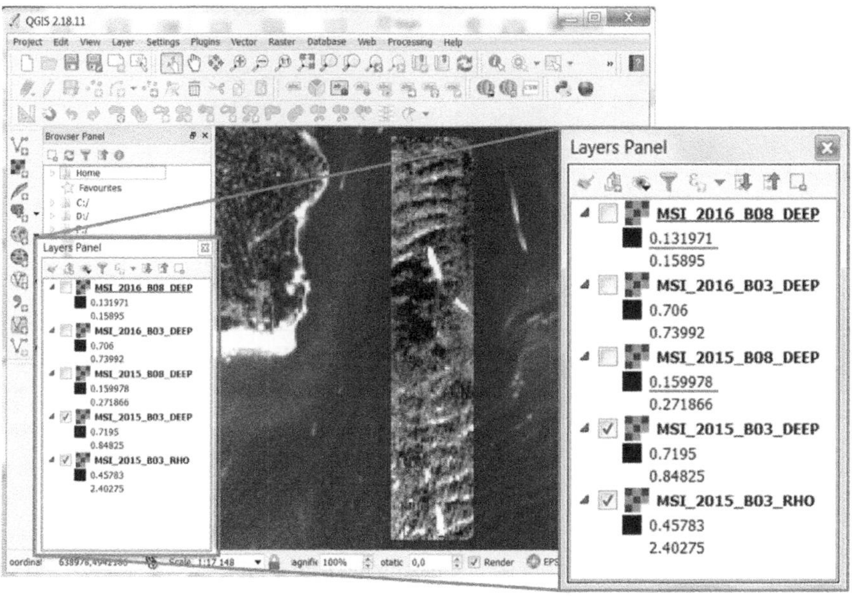

Figure 1.7. *Information displayed in the Layers Panel. The minimum value of* $\rho_{TOA}(NIR)$ *for the MSI_2015_B08_DEEP.TIF and MSI_2016_B08_DEEP.TIF layers are underlined in red. For a color version of the figure, see www.iste.co.uk/baghdadi/qgis4.zip*

	Determination of ***b*** values for 2015 and 2016 (Figure 1.8) • In the **Menu Bar** : - Select **Processing > Toolbox** - Select the module **GRASS GIS 7 commands > Raster > r.regression.line**

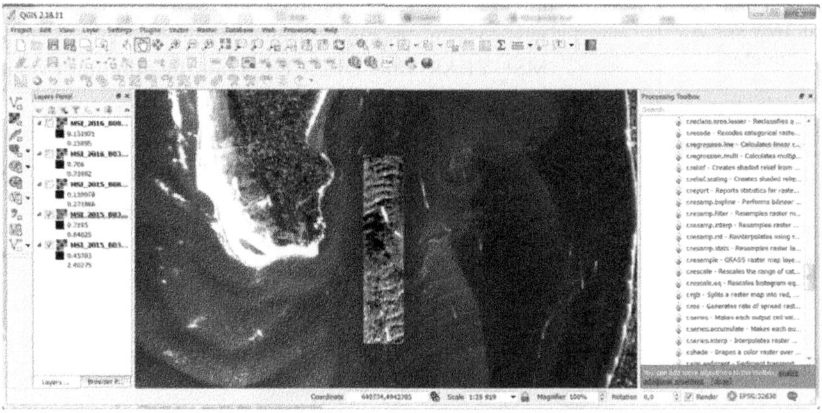

Figure 1.8. *Illustration of the processing possibilities offered by the QGIS Toolbox*

	• In the **R.Regression.Line Dialog Box**: - Select **Parameters** - **Layer for x coefficient**: Select .../MSI_YEAR_B08_DEEP.TIF - **Layer for y coefficient**: Select .../MSI_YEAR_B03_DEEP.TIF - Click on the **Run** button - Read the value of b *Example of b values: 1.1058 for 2015 and –0.0168[6] for 2016 (Figure 1.9).*

6 The sun glint impact is not significant in 2016. A value of b close to zero indicates no correlation between $\rho_{TOA}(NIR)$ and $\rho_{TOA}(G)$. This can be interpreted as the result of homogeneous and low-intensity environmental noise over the whole study area, which is assumed optically homogeneous.

<table>
<tr><td colspan="2">
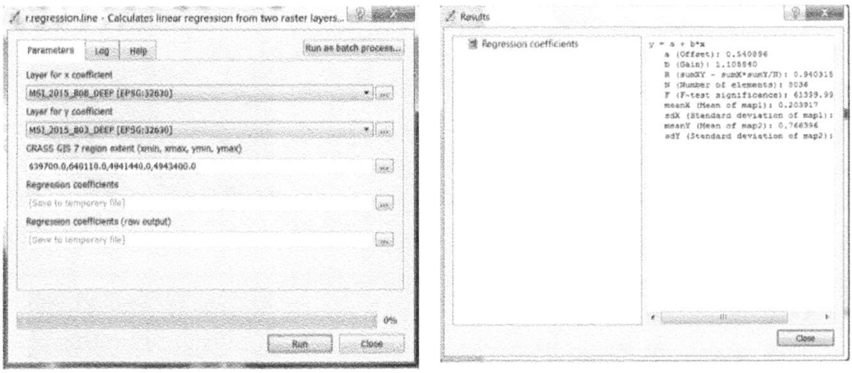

Figure 1.9. *Illustration of the QGIS processing used to perform a linear regression with the module r.regression.line*
</td></tr>
<tr><td>2.3. Application of the Hedley et al. [HED 05] model</td><td>
In the Menu Bar:
• Select Raster > Raster calculator…
In the Raster Calculator Dialog Box:
• Raster calculator expression :
"MSI_YEAR_B03_RHO@1" +b*(MIN-"MSI_YEAR_B08_RHO@1")
where b = 1.1058 and MIN = 0.1600 in 2015; b = -0.0168 and MIN = 0.1320 in 2016.
• Output layer :
- Select …/MSI_YEAR_B03_COR.TIF
See Figure 1.10 for the QGIS processing illustration.
</td></tr>
</table>

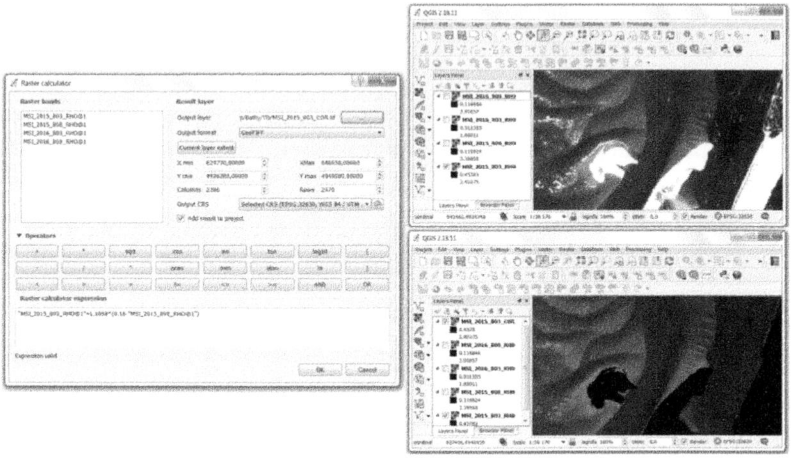

Figure 1.10. *On the left, illustration of the QGIS processing used to apply the correction of Hedley et al. [HED 05]; on the top right, the initial image of* $\rho_{TOA}(G)$*; on the bottom right, result of the Hedley et al. [HED 05] correction*

1.3.3. *Step 2: calculation of bathymetry*

Here, the objective is to map the bathymetry in the Arcachon inlet and adjacent shorelines using the empirical model described by equation [1.4]. The field bathymetry data used to calibrate the model were corrected beforehand. These data are distributed as shapefiles called Bathy_2015.shp and Bathy_2016.shp. The first step is to convert these vector data to the raster format.

Step	QGIS processing
1. Conversion of the field bathymetry data	In **QGIS**: • Open the following shapefiles: - Bathy_YEAR.shp In the **Menu Bar**: • Select **Raster > Conversion > Rasterize (Vector to Raster)...** In the **Rasterize Dialog Box**: • Input file : - Select successively the two shapefiles

- Attributed field :
  - Select bathy
- Output file :
  - Select …/BATHY_YEAR.TIF
  - A window "Output size or resolution required" opens automatically with the message "The output file doesn't exist…"
  - Click on the **OK** button
- Click on the **OK** button

See Figure 1.11 for the processing illustration.

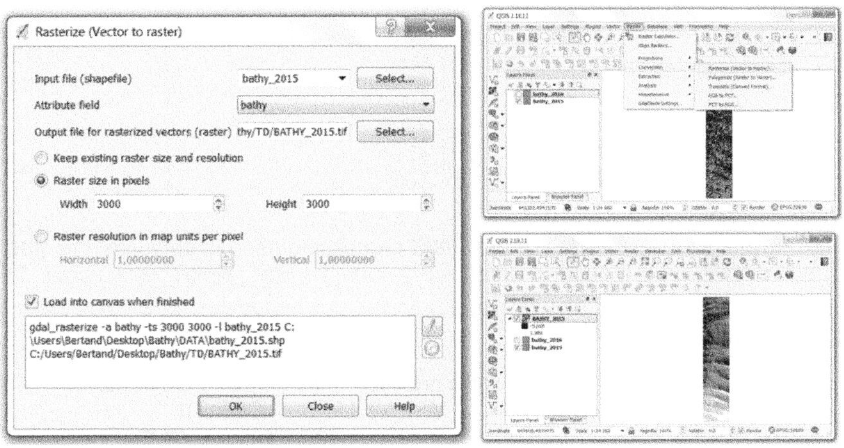

Figure 1.11. *On the left, illustration of the QGIS processing used to rasterize vector data. At the top right, field bathymetry data stored in a vector format; at the bottom right, result of the raster conversion. For a color version of the figure, see www.iste.co.uk/baghdadi/qgis4.zip*

2. Extraction of $\rho_{TOA}^{deep}(G)$

In **QGIS**:

- Check the presence of the following raster files:
  - MSI_YEAR_B03_DEEP.TIF

In the **Menu Bar**:

- Select **Processing** > **Toolbox**
  - Select the module **GRASS GIS 7 commands** > **Raster** > **r.quantile**
- In the **R.Quantile Dialog Box**:
  - Select **Parameters**
  - **Input raster layer** :
    Select …/ MSI_YEAR_B03_DEEP.TIF

<table>
<tr><td></td><td>

- **Number of quantiles**:
Enter the value 4
- Click on the button **Run**
- Read the value associated with the first quantile

Example of first quantile values: 0.7450 in 2015 and 0.7170 in 2016 (Figure 1.12).

</td></tr>
<tr><td colspan="2">

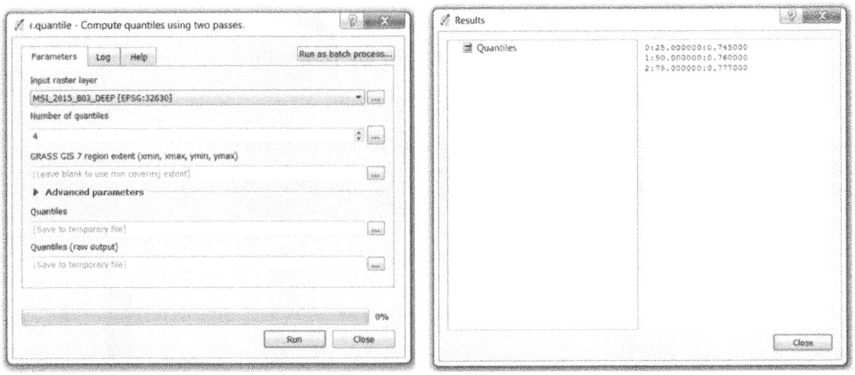

Figure 1.12. *Illustration of the QGIS processing used to calculate the first quantile values associated with the ρ_{TOA} images MSI_YEAR_B03_DEEP.TIF*

</td></tr>
<tr><td>3. Determination of the α and β parameters [1.4]</td><td>

In **QGIS**:

- Check the presence of the following raster files:
  - MSI_YEAR_B03_COR.TIF

In the **Menu Bar**:

- Select **Raster > Raster calculator…**

In the **Raster Calculator Dialog Box**:

- Raster calculator expression:

ln("MSI_2015_B03_COR@1"-0.7450)
or
ln("MSI_2016_B03_COR@1"-0.7170)

- Output layer:
  - Select .../ MSI_YEAR_X.TIF

In the **Menu Bar**:

- Select **Processing > Toolbox**
- Select the module **GRASS GIS 7 commands > Raster > r.regression.line**

</td></tr>
</table>

<table>
<tr>
<td></td>
<td>

- In the **R.Regression.Line Dialog Box**:
  - Select **Parameters**
  - **Layer for x coefficient**:
Select .../MSI_YEAR_X.TIF
  - **Layer for y coefficient**:
Select .../ BATHY_YEAR.TIF
  - **Grass GIS 7 region extent**
Select the option "Use layer/canvas extent"
Enter BATHY_YEAR.TIF
Click on the button **Run**
  - Read the value of α and β.

Example: $\alpha = 3.7209$ *and* $\beta = 2.2593$ *for 2015;* $\alpha = 5.0817$ *and* $\beta = 3.2804$ *for 2016 (Figure 1.13).*

</td>
</tr>
<tr>
<td colspan="2">

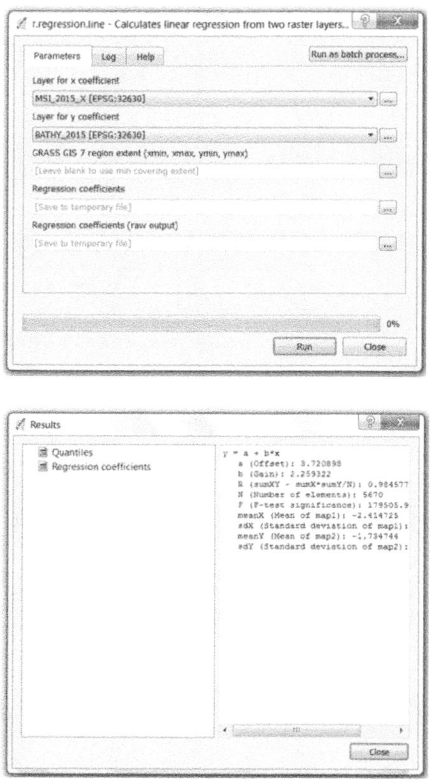

Figure 1.13. *Illustration of the QGIS processing used to perform a linear regression between the MSI_YEAR_X.TIF and BATHY_YEAR.TIF variables using the module r.regression.line*

</td>
</tr>
</table>

4. Bathymetry mapping	In the **Menu Bar**: • Select **Raster > Raster calculator…** In the **Raster Calculator diaog Box**: • Raster calculator expression: $\alpha + \beta \times$ `"MSI_YEAR_X@1"` *where* α*=3.7209 and* β*=2.2593 in 2015;* α*=5.0817 and* β*=3.2804 in 2016.* • Output layer: - Select `.../MSI_YEAR_BATHY.TIF` See *Figure 1.14 for the illustration of the processing.* 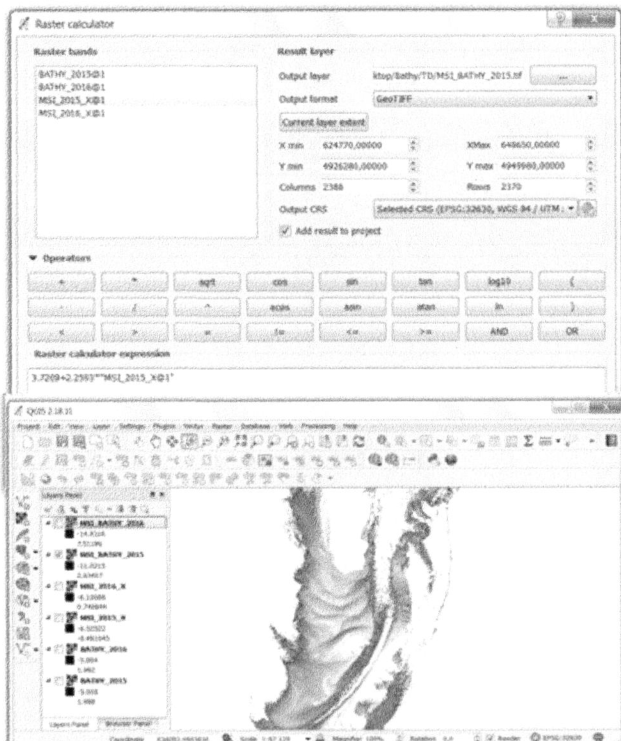

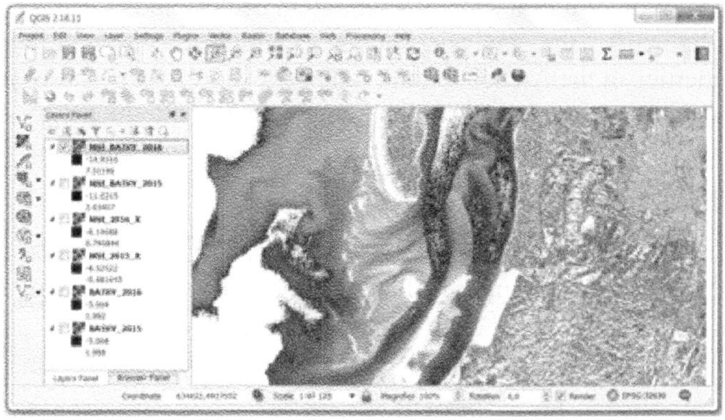

Figure 1.14. *On the top, illustration of the QGIS processing used to perform the bathymetry mapping over the whole area of the Arcachon inlet; middle, bathymetry map of 2015; bottom, bathymetry map of 2016*

In Figure 1.14, it is important to note two results that need to be corrected: (1) bathymetry was calculated for aquatic surfaces as well as continental/emerged surfaces and (2) bathymetry was calculated over a wide range of values that vary from –14.9 m to 7.5 m. The next step will therefore be dedicated to masking pixels associated with non-aquatic surfaces as well as bathymetry values that are outside the validity range.

1.3.4. *Step 3: preparation and application of masks*

Masking pixels with false bathymetry values is a major step. Here, the objective is to mask pixels of bathymetry maps that are associated with non-aquatic surfaces as well as bathymetry values that are outside the validity range. The mask "continental surface" is based on the use of a threshold value of ρ_{TOA} in the SWIR band (channel 11), which allows the classification of continental and aquatic surfaces. It is important to note that the images acquired in the SWIR band have a spatial resolution of 20 m (see Table 1.2). It is therefore required to resample these images to 10 m using an interpolation method.

The mask "*H* not valid" is based on the use of a lower and upper bound value of the bathymetry which define the interval of validity of the retrieval. These values are determined from a qualitative analysis of the `MSI_YEAR_B03_COR.TIF` and `MSI_YEAR_BATHY.TIF` images. This analysis shows that sub-marine sedimentary structures are well delimited for high water depths, which supports the assumption of a low vertical attenuation of light in the water column (see Table 1.1). Additionally, the analysis shows a high (low) spatial homogeneity of optical properties in the Arcachon inlet (adjacent shorelines). These good conditions make it possible to extend the validity range of the bathymetry product defined by Capo *et al.* [CAP 14]. For this exercise, I propose the following lower (Z_{inf}) and upper (Z_{sup}) bounds:

– limit value for low water depths: $H_{inf} = -0.5$ m → $Z_{inf} = T - 0.5$ m;

– limit value for intermediate water depths: $H_{sup} = -5$ m → $Z_{sup} = T - 5$ m.

<table>
<tr><th>Step</th><th>QGIS processing</th></tr>
<tr><td>1. Resampling of SWIR images at 10 m</td><td>In QGIS:
• Check the presence of the following raster files:
- MSI_YEAR_B11_RHO.TIF
In the Menu Bar:
• Select Processing > Toolbox
• Select the module GRASS GIS 7 commands > Raster > r.resamp.interp
• In the R.Resamp.Interp Dialog Box:
- Select Parameters
- Input raster layer:
Select .../MSI_YEAR_B11_RHO.TIF
- Sampling interpolation method
Select the option “bilinear”
- GRASS GIS 7 region cell size
Enter the value 10
- Resampled interpolated > Save to file
Select .../MSI_YEAR_B11_INT.TIF
- Click on the button Run
See Figure 1.15 for the processing illustration.</td></tr>
</table>

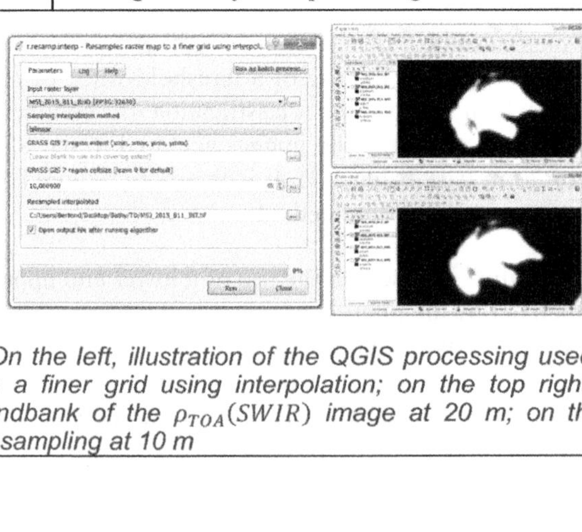

Figure 1.15. *On the left, illustration of the QGIS processing used to resample a raster map to a finer grid using interpolation; on the top right, zoom on the Toulinguet sandbank of the* $\rho_{TOA}(SWIR)$ *image at 20 m; on the bottom right, result of the resampling at 10 m*

<table>
<tr>
<td>2. “Continental surface” mask</td>
<td>
Determination of the threshold value (Figure 1.16)
• In the Layers Panel:
- Select the MSI_YEAR_B11_INT.TIF layer
- Select Properties > Histogram
- Read the threshold value (the upper bound of the first mode for which the frequency becomes quasi-zero)
Example of threshold values: 0.2 in 2015 and 0.1 in 2016.
Thresholding (Figure 1.17)
• In the Menu Bar:
- Select Raster > Raster calculator...
• In the Raster Calculator Dialog Box:
- Raster calculator expression:
"MSI_YEAR_B11_INT@1"<Threshold_Value
where Threshold_Value = 0.2 in 2015 and 0.1 in 2016.
• Output layer:
- Select .../MASK_YEAR_SE.TIF
Encoding of the “no data values” (ND) (Figure 1.18)
• In the Menu Bar:
- Select Raster > Projections > Warp (Reproject)...
- Input file
Select .../MASK_YEAR_SE.TIF
- Output file
Select .../MASK_YEAR_SE_ND.TIF
- Select the option “No data values”
Enter the value 0
- Click on the button OK
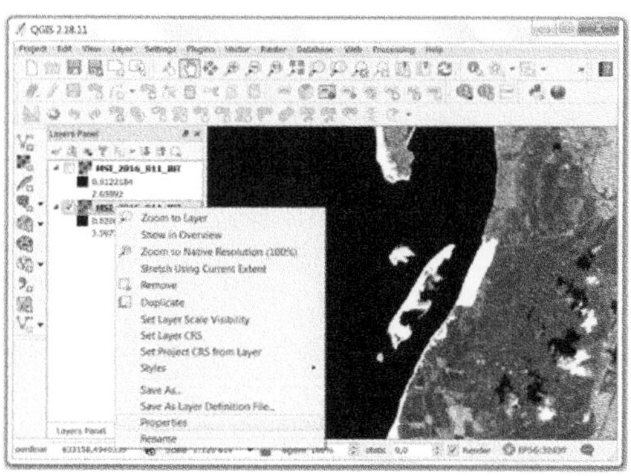

</td>
</tr>
</table>

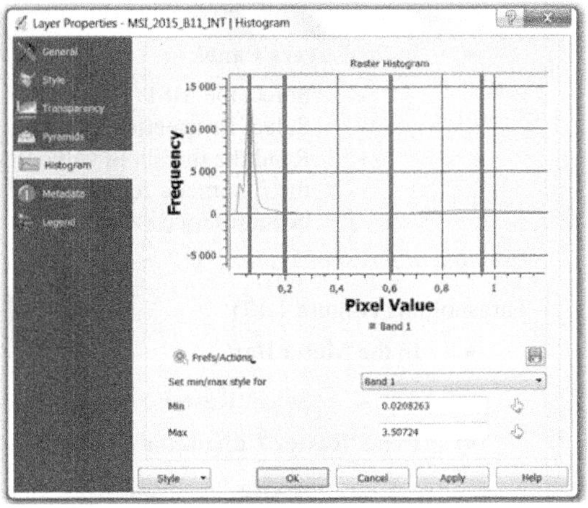

Figure 1.16. *Above, illustration of the QGIS processing used to determine a threshold value from the analysis of the frequency histogram. Below, histogram of the* $\rho_{TOA}(SWIR)$ *values of the image acquired in 2015. Two modes can be observed with a central value at approximately 0.06 and 0.95 (located by a vertical green line). The threshold value for the first mode can take any value between 0.2 and 0.6 (located by a vertical red line), without affecting the overall quality of the mask. The lower value (0.2) is selected by default. For a color version of the figure, see www.iste.co.uk/baghdadi/qgis4.zip*

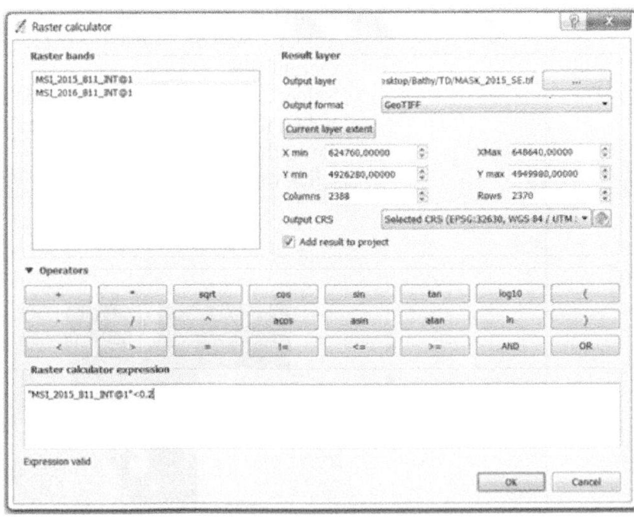

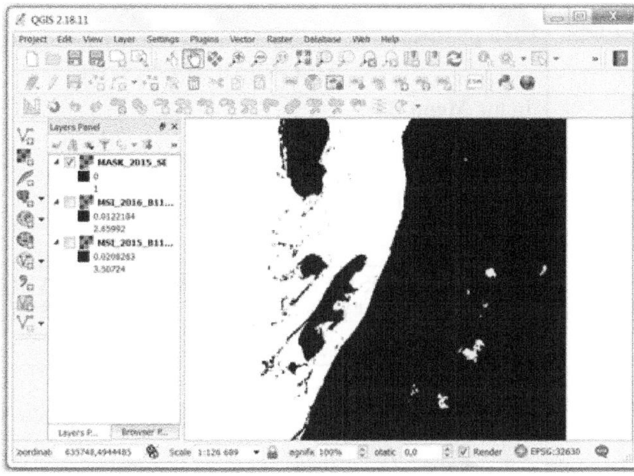

Figure 1.17. *Above, illustration of the QGIS processing used to perform a thresholding; below, the mask "continental surface" where aquatic (continental) pixels take the value 1 (0)*

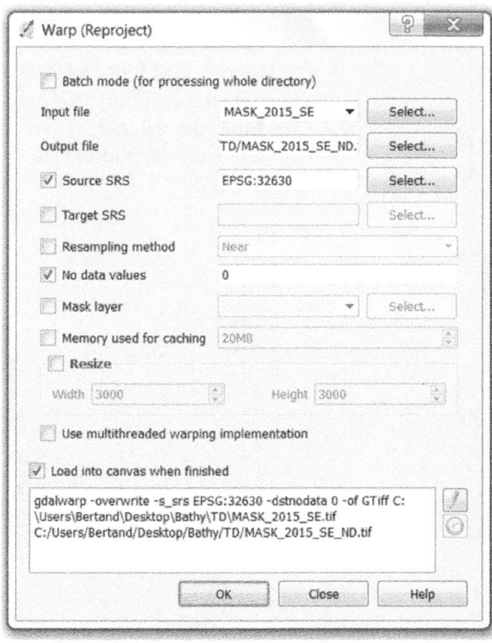

Figure 1.18. *Illustration of the QGIS processing used to encode no data values*

3. "*H* not valid" mask	In **QGIS**: • Check the presence of the following rater files: - MSI_YEAR_BATHY.TIF In the **Menu Bar**: • Select **Raster > Raster calculator...** • In the **Raster Calculator Dialog Box**: - Raster calculator expression: "MSI_YEAR_BATHY@1"<T-0.5 AND "MSI_YEAR_BATHY@1">T-5 *where T = 2.14 m in 2015 and 2.44 m in 2016.* • Output layer: - Select .../MASQUE_YEAR_NV.TIF *See Figure 1.19 for the processing illustration.* In the **Menu Bar**: - Select **Raster > Projections > > Warp (Reproject)...** - Input file Select .../MASK_YEAR_NV.TIF - Output file Select .../MASK_YEAR_NV_ND.TIF - Select the option "No data values" Enter the value 0 - Click on the button **OK**

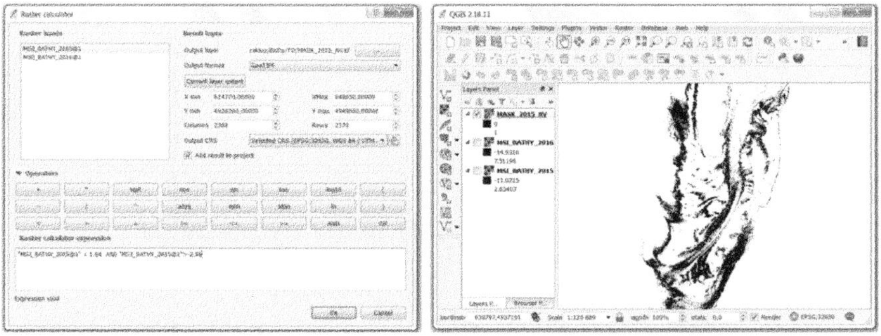

Figure 1.19. *On the left, illustration of the QGIS processing used for the thresholding; on the right, the mask "H not valid" where pixels take the value 1 if –2.86 < Z < 1.64, 0 otherwise*

4. Merging of the masks and application	In the **Menu Bar**: • Select **Raster > Raster calculator...** • In the **Raster Calculator Dialog Box**: - Raster calculator expression: "MSI_BATHY_YEAR@1"*"MASQUE_YEAR_NV_ND@1" *"MASQUE_YEAR_SE_ND@1" • Output layer: - Select .../MSI_BATHY_YEAR_MASK.TIF *See Figure 1.20 for the processing illustration.*

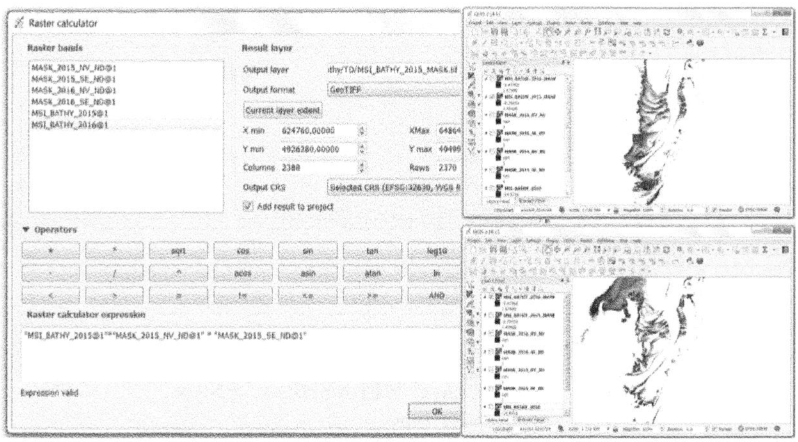

Figure 1.20. *On the left, illustration of the QGIS processing used to apply a mask on a bathymetry map; on the top right, bathymetry map of 2015; on the bottom right, 2016 bathymetry map. An extension of the bathymetry off the Cap Ferret sand spit can be observed in 2016. This is an artifact due to a high resuspension in this area*

1.3.5. *Step 4: characterization of the morphological evolution of the main submarine sedimentary structures*

Step	QGIS processing
1. Estimation of the variations of bathymetry	In **QGIS**: • Check the presence of the following raster files: - MSI_BATHY_YEAR_MASK.TIF

<table>
<tr><td></td><td>In the Menu Bar:

• Select Raster > Raster calculator…
• In the Raster Calculator Dialog Box:

- Raster calculator expression:

"MSI_BATHY_2016_MASK@1"-
"MSI_BATHY_2015_MASK@1"

• Output layer:

- Select .../DIFF_BATHY.TIF
See Figure 1.21 for the processing illustration.</td></tr>
<tr><td colspan="2">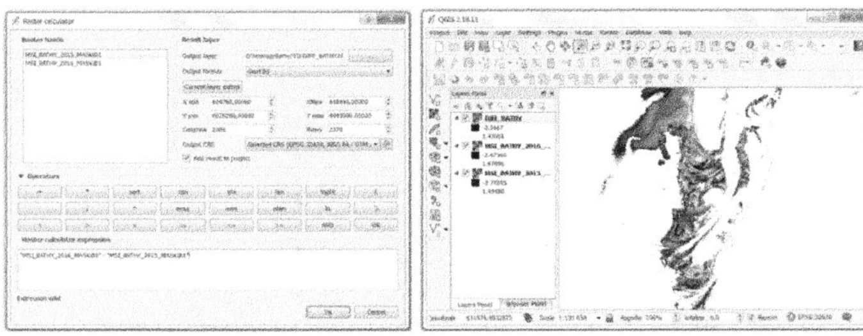

Figure 1.21. Illustration of the QGIS processing used to estimate the variations of bathymetry</td></tr>
<tr><td>2. Formatting of results</td><td>In the Menu Bar:

• Select Layer > Properties… > Style
- Render type
Select the option “Singleband pseudocolor”
- Color
Select your color
- Click on the button Edit
- Click on the button Apply

In the Menu Bar:

• Select Project > New Print Composer
- Give a name to your project

In the Composer Dialog Box:

• Click on the button Add new map 
• Click on the button Add new legend 
• Click on the button Add new scalebar 
• Select Composer > Export as Image…

See Figure 1.22 for the processing illustration.</td></tr>
</table>

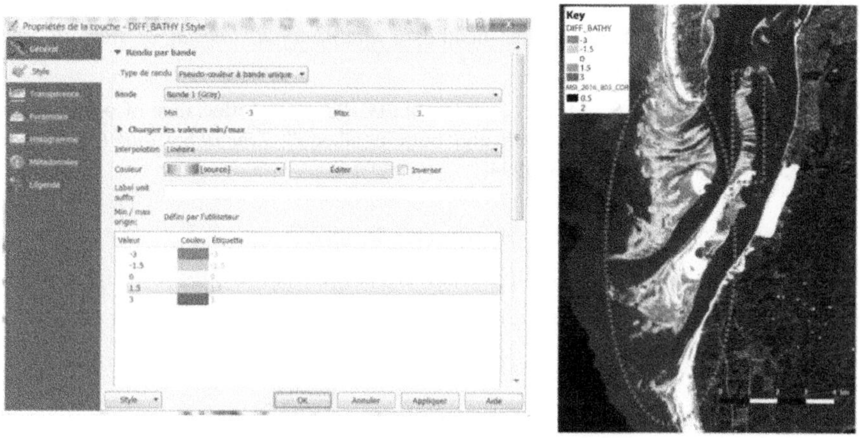

Figure 1.22. *Top: illustration of the QGIS processing used for the formatting of results; bottom: map of the bathymetry variations in the Arcachon inlet where the outer part (hydraulic dunes) is delimited by a red (green) contour. For a color version of the figure, see www.iste.co.uk/baghdadi/qgis4.zip*

Results provided by the analysis of Figure 1.22 could be compared to the results of Capo *et al.* [CAP 14]. Initially, the reader may focus their attention on two points in order to characterize the general evolution of the morphology of the large submarine sedimentary structures:

– the large-scale behavior (O ~1 km) of the North-South transport of sediment by studying the succession of accumulation/erosion zones in the outer part of the Arcachon inlet and adjacent coasts;

– the medium-scale behavior (O ~0.1 km) of the hydraulic dunes located in the north of the Arguin bank[7] and characterized by a north migration.

1.4. Bibliography

[BRU 15] BRU D., Corrections atmosphériques pour capteurs à très haute résolution spatiale en zone littorale, PhD Thesis, Bordeaux University, France, 2015.

[CAP 14] CAPO S., LUBAC B., MARIEU V. *et al.*, "Assessment of the decadal morphodynamic evolution of a mixed energy inlet using ocean color remote sensing", *Ocean Dynamics*, vol. 64, pp. 1517–1530, 2014.

7 The Arguin bank (France) is the emerged sand bank located between the north and south channel.

[HED 05] HEDLEY J.D., HARBONE A.R., MUMBY P.J., "Simple and robust removal of sun glint for mapping shallow-water benthos", *International Journal of Remote Sensing*, vol. 26, pp. 2107–2112, 2005.

[MAR 94] MARITORENA S., MOREL A., GENTILI B., "Diffuse reflectance of oceanic shallow waters: influence of water depth and bottom albedo", *Limnology and Oceanography*, vol. 39, no. 7, pp. 1689–1703, 1994.

2

Contribution of the Integrated Topo-bathymetric Model for Coastal Wetland Evolution: Case of Geomorphologic and Biological Evolution of Ichkeul Marshes (North Tunisia)

2.1. Coastal wetland dynamic

Coastal wetlands are among the most productive ecosystems. They provide many important services to human society. However, they are also ecologically sensitive. Wetlands exhibit enormous diversity according to their geographical location, water regime, dominant species, soil and sediment characteristics.

In wetland areas, there are strong interactions between water, soil, vegetation and human activities. Coastal vegetation species and communities are mainly dependent on the topographic-bathymetry level for both submerged and emergent areas. Mapping vegetation communities is an effective tool for coastal monitoring. This can help us determine the actual state of wetland areas in order to create management strategies, e.g. drawing the boundaries for protection zones in coastal wetland ecosystems.

Chapter written by Zeineb KASSOUK, Zohra LILI-CHABAANE, Benoit DEFFONTAINES, Mohammad EL HAJJ and Nicolas BAGHDADI.

Remote sensing and the Geographical Information System (GIS) have been useful in this area due the capability of satellite remote sensing to provide synoptic, repetitive and multispectral images, and due to the multivariate analysis offered by GIS. Remote sensing derived data have proved to be useful in various coastal wetland studies, such as for vegetation mapping and monitoring, change detection, biotope mapping, shoreline changes and suspended sediment dynamics systems.

This chapter, therefore, focuses on the reliability of remote sensing imagery combined with topographic and bathymetric data to map the coastal wetland communities in Ichkeul marshes. Marshes are a part of the Ichkeul park which is inscribed as important wetlands on the mediterranean coast.

2.2. Ichkeul marshes wetland

Ichkeul wetland is located in northern Tunisia (Figure 2.1). It covers an area of about 12,500 ha divided into three major geomorphological units: the brackish water lake (8,500 ha), the marshes (between 2,737 and 3,500 ha) and the mountain massif, also known as Djebel Ichkeul (1,363 ha). It is a UNESCO Biosphere Reserve (1977), a World Heritage site (1979), a National Park and a Ramsar site (1980). Despite its protected status, Ichkeul wetland is changing due to anthropogenic pressure in the watershed area, such as agriculture and the damming of the five main inflowing rivers (Oueds). Historically, the lake water level fluctuated on an annual cycle: from high water levels during the winter and the spring when fresh water flows out through the Tinja channel, to low water levels in summer when most of the inflowing rivers are dry and evaporation prevails. Marine water then flows into the lake and up to the channel. Alterations to this natural water level cycle through regulation are well documented and known to affect wetland community dynamics. Increased evaporation and evapotranspiration is leading to decreased soil moisture, reduced lake fertility and enhanced salinity [KAS 08, TAM 94].

Figure 2.2 is an image of a wetland landscape with Joumine marsh, the lake and the mountain (Djebel) in the background.

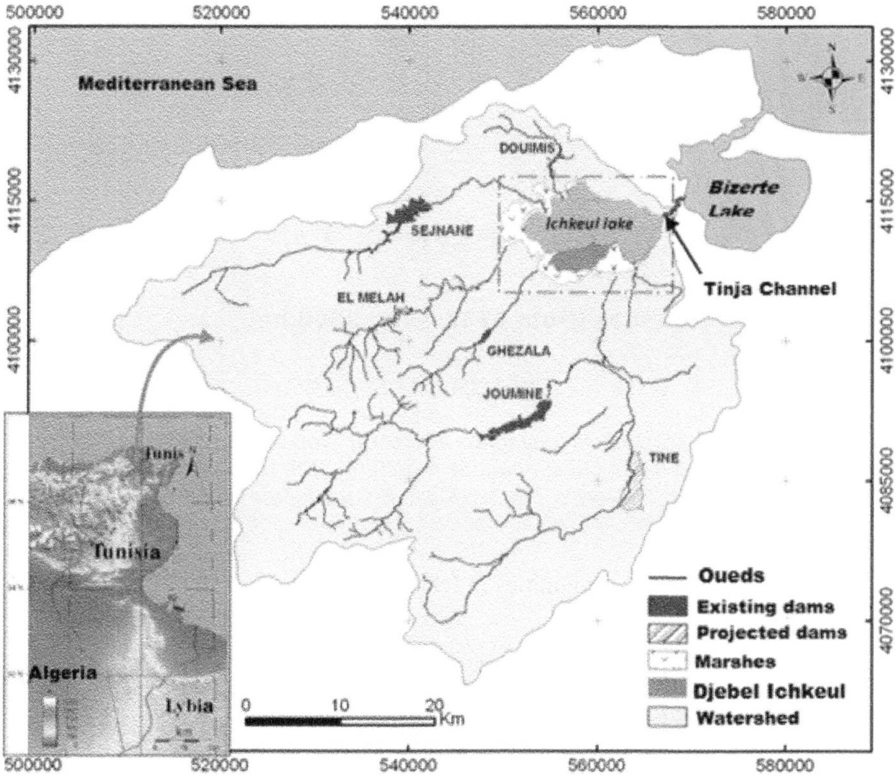

Geographic system : WGS84/UTM zone 32 N

Figure 2.1. *Location of Ichkeul ecosystem in northern Tunisia (Latitude 37°09′52″ N, Longitude 09°40′07″E). The interconnection of the Ichkeul Lake with the sea is made by the Tinja channel which leads into the marine Bizerte Lake. The Ichkeul ecosystem includes the wooded massif (Djebel), the lake, the six wetland marshes and the adjacent uplands. For a color version of the figure, see www.iste.co.uk/ baghdadi/qgis4.zip*

Figure 2.2. *Typical landscape of Ichkeul wetland (the picture contains the Ichkeul Lake, Joumine marsh and the Djebel). For a color version of the figure, see www.iste.co.uk/baghdadi/qgis4.zip*

Dominant vegetation communities in Ichkeul marshes are:

– Flooded vegetation:

- *Picris echioides* community: composed of *Picris echioides, Sylibum marianum* and *Galactites tomentosa*;

- *Ammi visnaga* and *Galactites tomentosa* community: characterized by the presence of *Galactites tomentosa, Ormenis praecox, Ridolfia segetum* and *Cirsium syriacum, Ammi visnaga, Capnophyllum peregimum, Sylibum marianum*;

- *Fustuca eliator ssp. arundincea* and *d'Oenanthe globulosa* community that are developed across flooded plains.

– Halophilic communities:

- *Hordeum maritimum ssp. Eumaritimum* (Hm) community: this community indicates a ground low-lying alkaline area and poor soil structure;

- Hygrophilous community (Hy): this community is characterized by *Juncus subulatus, Juncus maritimus, Scirpus maritimus, Phragmites communis, Thypha angustifolia, Scirpus littoralis*. Plant species within this community are tolerant to prolonged flooding (more than six months) and soil salinity. *Frankenia thymifolia* and *Hordeum maritimum* are developed in less submerged soil.

2.3. Object-oriented classification method integrating the topo-bathymetric terrain model

Previous approaches to the use of remote sensing data for wetland vegetation mapping have often focused on study sites with relatively limited species diversity or have attempted to only differentiate broad vegetation communities. The previous mapping of Ichkeul vegetation was based on either field observations and photo-interpretation, or pixel classification of remote sensing data [GHR 06]. However, the methods used do not allow good delineations of Ichkeul marsh's vegetation communities. In contrast, the object-oriented approaches outperformed pixel-based classification methods.

In this study, we will combine both the object-oriented classification of high spatial resolution imagery and the integrated topographic-bathymetric digital terrain model (topo-bathymetric DTM hereafter) in order to characterize and to map vegetation communities across Ichkeul marshes.

The chapter is divided into two parts:

– first, we settle the integrated topo-bathymetric DTM by combining three datasets: the lake bathymetry, the marshes and the mountain digital topographies;

– second, in order to classify Ichkeul marshes, the new topo-bathymetric DTM is used with multispectral imagery from Advanced Spaceborne Thermal Emission and Reflection Radiometer (ASTER) images from the Terra satellite (Table 2.1). Classification is carried out using the object-oriented process based on botanical field observations.

The method developed in this chapter uses the free and open-source software GIS QGIS (version 2.18.3) and the GRASS GIS plugin (version 7.2.0). This software is used for geospatial data management and analysis, image processing, graphics and maps production, spatial modeling and visualization. It runs on many operating systems such as Linux, Microsoft Windows, Mac OSX and Android.

Figure 2.3 illustrates the sequence of treatments required for the application of the proposed method. To facilitate its reading, the treatments are grouped into seven main stages:

– building of the topo-bathymetric DTM;

– radiometric correction of ASTER image with a focus on the study area;

– calculation of the Normalized Difference Vegetation Index (NDVI);

– principal component analysis (PCA);

– image segmentation (segmented image = stack of the NDVI, the first PCA component and the DTM);

– identification of training samples in the segmented image and calculation of their spectral and thematic attributes;

– classification of the segmented image.

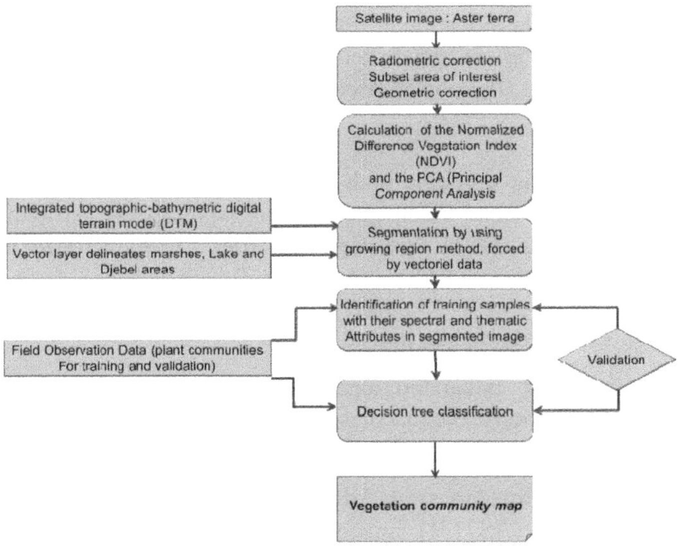

Figure 2.3. *Flow chart for vegetation community mapping combing remote sensing data, field measurement and topo-bathymetric DTM*

2.3.1. *Construction of the topo-bathymetric DTM*

Many biological applications in coastal environments require knowledge of the near-shore topography and bathymetry. However, existing topographic and bathymetric data are collected independently for different purposes; it has been difficult to use them together at the land–water interface due to differences in format, projection, resolution and accuracy. Three input data form the basis for the computation of the topo-bathymetric DTM, consisting of points and lines (contours):

– bathymetric data are obtained from echo-sounding campaigns recommended by the conservation authority of the Ichkeul Park in 2003, focusing on the lake and the Tinja channel. The bathymetric dataset consists of 220,000 measurements, with an average measurement density of 2.5 points per m^2;

– a topographic survey of the marsh areas done in 2003 (vector layer in point format) with a planimetric mesh varying from 2 to 4 m;

– data covering Djebel Ichkeul and the surrounding lands has been obtained by digitalizing four 1:25,000 scale topographic maps (produced in 1995) covering the Ichkeul park (contour lines and hilltops).

The quality of the produced integrated topographic/bathymetric DTM depends on the spatial variation of the accuracy and the density of the altitudinal data. Accuracy may be measured in height or in horizontal position (e.g. for breaklines).

In this chapter, we first merge topographic and bathymetric data together in order to have only one dataset format. Contour lines of topographic data are converted into points. Line vertices become points with the height attribute. For this, we have used the extract nodes function installed in QGIS.

QGIS functionality to extract nodes from contour lines: - Vector > Geometry tools > Extract nodes
QGIS functionality to merge many vector layers: - Vector > Vector general Tools > Merge vector layers

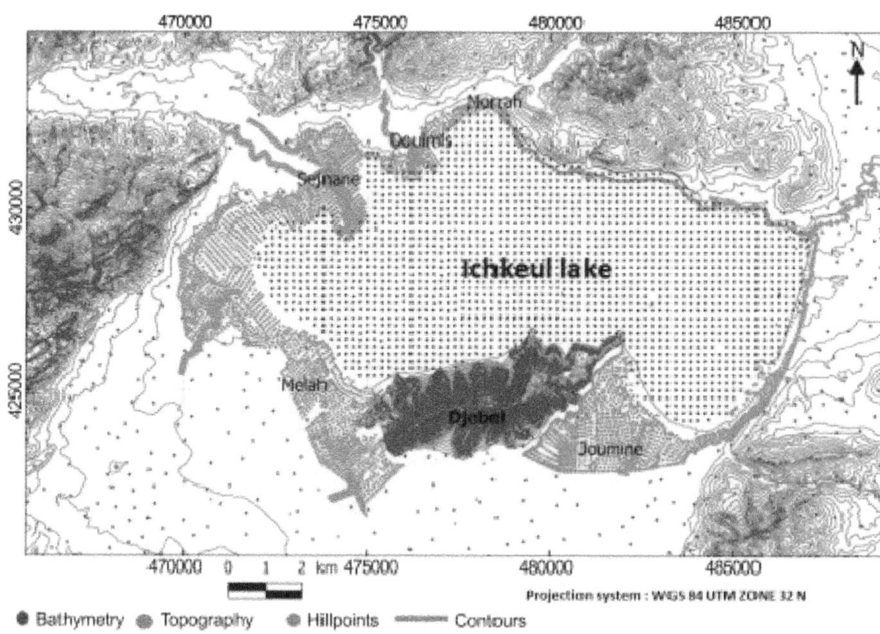

Figure 2.4. *Map with bathymetric (for the lake) and topographic data for Djebel and marshes (the red lines and dots show the topography of Djebel and the lands around the park, yellow dots represent point topographic surveys in marshes and blue dots are the bathymetric measurements in the lake). For a color version of the figure, see www.iste.co.uk/baghdadi/qgis4.zip*

The second step is to calculate the map density. Density can be understood as the number of points per unit area or as a pixel value that corresponds to the number of points within the area covered by the pixel. Data density solely depends on the horizontal positions of the data. A density map depicts the amount of discretization of the DTM. Areas covered by few or no data become distinguishable. The completeness of the altitudinal data may be inspected, as density is equal to zero in data voids.

The QGIS core plugin called "Heatmap" is used to create heat maps from altitudinal point data. This plugin needs to be activated first. Heatmaps are used to easily identify where there is a high concentration of observations [VEN 02].

QGIS Functionality to create a Heatmap:
- Install the "Heatmap" plugin: Plugins > Manage and Install Plugins
- Density map: Raster > Heatmap

In the study area, the maximum point density is 4 points/100 m². The highest density of points is in marsh areas and on the lake boundary. The high and homogeneous point density in those areas enabled the making of a detailed analysis of the altitude change in marshes. In the areas shown in white there are no points. The lowest density is observed in hill zones (Figure 2.5).

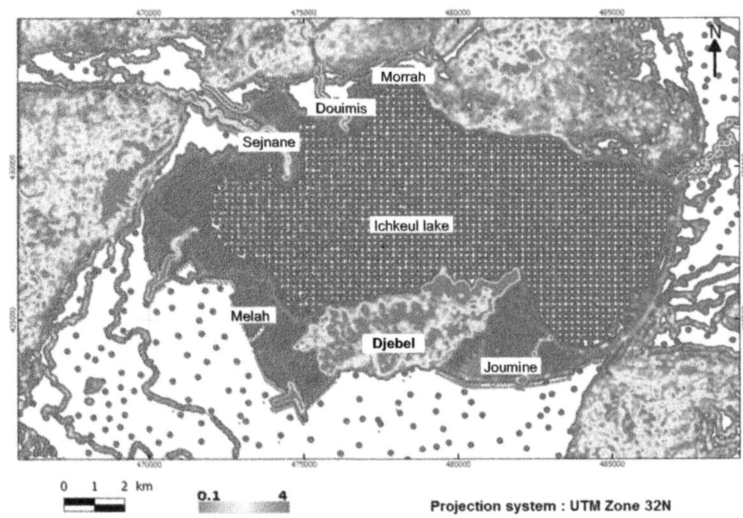

Figure 2.5. *Density map of topographic and bathymetric data, computed with a cell size of 100 m x 100 m. Areas with high density are observable. Cells without data are colored in white. For a color version of the figure, see www.iste.co.uk/baghdadi/qgis4.zip*

Point elevation data covering the emerged and submerged areas are then interpolated using the Triangular Irregular Network (TIN) algorithm of QGIS® software. The result consists of a grid digital elevation model with a resolution of 15 m × 15 m (Figure 2.6). In this study, the DTM and the used satellite data have the same ground resolution.

QGIS functionality to create a DTM by interpolation
- Raster > Interpolation

The integrated topographic/bathymetric DTM (Figure 2.6) shows the overall morphology of the Ichkeul ecosystem that consists of three main geographic units, which are the lake, the marshes and the isolated wooded massif of the Djebel Ichkeul. The produced DTM is used to perform the classification of the marsh's vegetation. Only the marsh's units are studied here.

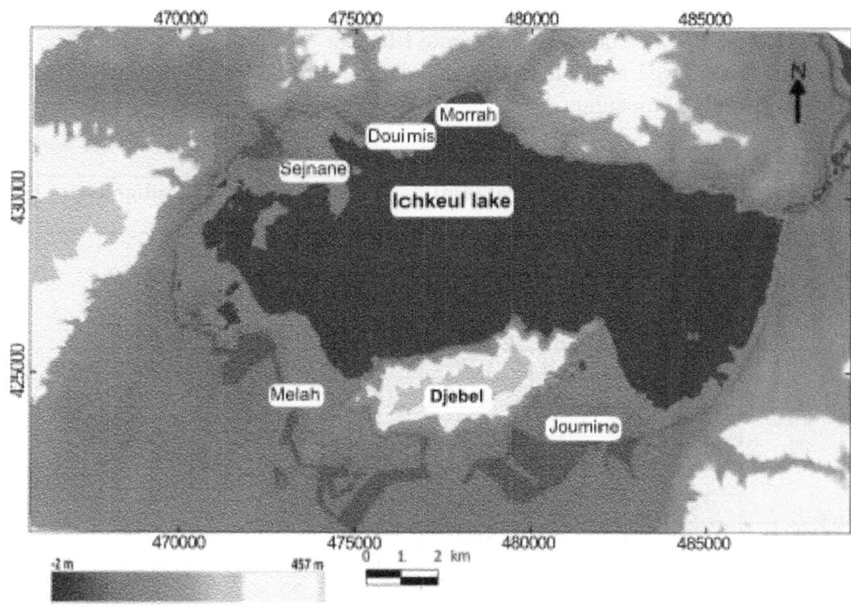

Figure 2.6. *Hillshade from the new integrated topographic bathymetric DTM of the Ichkeul ecosystem in grid format (pixel 15 m × 15 m). The elevation is represented in a meter relative to the Tunisian Altitudinal System (NGT). The coordinate system is UTM 32N. For a color version of the figure, see www.iste.co.uk/baghdadi/qgis4.zip*

2.3.2. Image preprocessing

2.3.2.1. *Import of the ASTER data image*

In order to emphasize the growing period of vegetation cover (May, June and July) and the constraints of access to data, we have used an ASTER image acquired on June 8th 2007. An overview of all the used spatial data is given in Table 2.1.

ASTER data include three visible and near infrared (VNIR) bands (with 15 m spatial resolution), six shortwave infrared (SWIR) bands (with 30 m spatial resolution) and five thermal infrared (TIR) bands (with 90 m spatial resolution). The used image is an ASTER Level-1B (L1B) Registered Radiance at the Sensor product. The L1B product contains radiometrically calibrated and geometrically coregistered data. ASTER products are downloaded via FTP and HTTPS (https://search.earthdata.nasa.gov).

The L1B ASTER product is two separate files: an "hdf" file and a "met" file. The hdf file contains a large amount of the metadata and the spectral data, whereas the met file only contains metadata.

Image	Resolutions			
	Bands	**Spectral (μm)**	**Spatial (m)**	**Radiometric**
VNIR	1	0.52–0.60	15	8 bits
	2	0.63–0.69		
	3N	0.78–0.86		
	3B	0.78–0.86		
SWIR	4	1.60–1.70	30	8 bits
	5	2.145–2.185		
	6	2.185–2.225		
	7	2.235–2.285		
	8	2.295–2.365		
	9	2.360–2.430		
TIR	10	8.125–8.475	90	12 bits
	11	8.475–8.825		
	12	8.925–9.275		
	13	10.25–10.95		
	14	10.95–11.65		

Table 2.1. *Characteristics of ASTER remote sensing data used for Ichkeul vegetation classification*

In this case study, only bands 1, 2 and 3N of the VNIR, and bands 4 and 5 of the SWIR (medium infrared) are used. The vegetation cover has specific characteristics in the red and near infrared domains. The spectral characteristics of hygrophilous vegetation are distinguished in the bands that highlight the water (very strong reflectance in blue and very low reflectance in the near infrared). This part of the visible region of the spectrum is often chosen, rather than using the entire visible region.

2.3.2.2. *Atmospheric correction*

The purpose of atmospheric correction is to produce more accurate surface reflectance and to improve the extraction of surface parameters from ASTER images. To achieve this, the influences of the atmosphere, solar illumination, sensor viewing geometry and terrain information have to be taken into account. Digital numbers (DN) in L1B are radiance at the sensor and are not reflectance on the ground. Therefore, in order to convert radiance (8 bits) at the sensor to reflectance (%) on the ground, we process an atmospheric correction. First, we convert DN into reflectance at the Top of the Atmosphere (TOA), and then we apply atmospheric correction to produce ASTER surface reflectance.

Conversion of the DN of the ASTER bands (1–10) to the TOA reflectance is carried out using a developed python script. The atmospheric correction is handled using the 6S atmospheric correction algorithm (Second Simulation of Satellite Signal in the Solar Spectrum).

The 6S model parameters (Table 2.2) for the bands of interest (bands 1, 2, 3N, 4 and 5) are provided in the chapter dataset.

6S parameters	**Value**
Geometrical conditions (depends on sensor, 10 for ASTER)	10
Month Day Hour (hh.ddd) Longitude ("hh.ddd" is in decimal hours GMT) Latitude ("hh.ddd" is in decimal hours GMT)	6 8 13.00 –47.410 –20.234
Atmospheric mode	1
Aerosols model: continental	1
Visibility [km] (aerosol model concentration)	15
Mean target elevation above sea level [km]	–0.600
Sensor height (here, sensor onboard a satellite) [km]	–1,000
Coefficient that corresponds to each band (72 for band 1)	72

Table 2.2. *Parameters of the algorithm 6S, associated with the sensor and the acquisition date of the ASTER image*

QGIS functionality to calculate the TOA reflectance:
- Conversion to TOA reflectance: Processing > Toolbox >Scripts > Add a script

GRASS functionality to convert DN values to surface reflectance:
- Processing > Toolbox > GRASS GIS 7 Commands > Imagery > i.atcorr

2.3.2.3. *Reprojecting and layer stacking*

In this study, we use the first five bands (1, 2 and 3N of VNIR images and four and five bands of SWIR). After atmospheric correction, the five bands are reprojected to the coordinate system UTM 32N. Finally, the reprojected bands are stacked and saved into a multiband image.

The next step is to clip the stacked image to the extent of the study area delimited by the polygon given in Table 2.3.

	X(E)	y (N)
Northwest corn	547910	4118570
Southeast corn	568790	4102520

Table 2.3. *Geographic coordinates (UTM 32N) of the study area*

QGIS Functionality to reproject a band:
- Raster > Projections > Warp (Reproject)

QGIS Functionality to stack bands:
- Raster > Miscellaneous > Build Virtual Raster

QGIS Functionality to clip a raster:
- Raster > Extraction > Clipper

2.3.2.4. *Calculation of the NDVI*

The NDVI (the normalized ratio of red and near infrared reflectance) has been widely used for vegetation detection [ROU 74]. Very low values of NDVI (0.1 and below) correspond to barren areas of rock, sand or snow. Moderate values represent shrub and grassland (0.2–0.3), while high values indicate temperate areas and rainforests (closest to 1). Bare soil is represented by NDVI values, which are closest to zero (0), and water bodies or immerged vegetation are represented by negative NDVI values. In this chapter, we use this ratio to map vegetation.

QGIS functionality for NDVI calculation:
- Raster > Raster calculator

2.3.2.5. *Principal component analysis*

The PCA is a statistical technique for transforming n-dimensional vectors (i.e. image bands). This process is done in order to compress spectral data and only retain the components that carry the most significant information.

Application of the PCA process to the first five ASTER images (1, 2, 3N, 4 and 5 bands) shows that the first three principal components (PC1, PC2 and PC3) contain the largest amount of information from the original dataset (99.5%) (Figure 2.7).

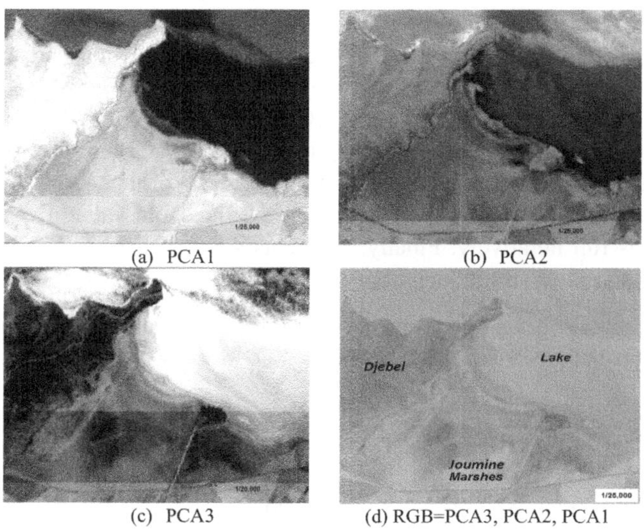

Figure 2.7. *Illustration taken on Joumine marsh (Figure 2.2), showing the principal component analysis results: a) PCA1, b) PCA2, c) PCA3 and d) RGB=PCA3, PCA2, PCA1. For a color version of the figure, see www.iste.co.uk/baghdadi/qgis4.zip*

The first principal component image PCA1 includes the largest percentage of the total scene variance (76.9%) (Figure 2.7(a)). This component enhances the topographical information, particularly the Djebel and the lake. The second principal component image PCA2 (Figure 2.7(b)) enhances the difference between vegetation areas and water bodies. The third principal component PCA3 (Figure 2.7(c)) enhances dense vegetation (blackness and darkness in the image). Trees of the forest in the Djebel and *Phragmites communis* vegetation on the lake are presented in dark.

A color composite of the first three principal components (R=3, G=2, B=1) where colors correspond to land cover classes (Figure 2.7(d)). Green represents water (lake) and vegetation, while red corresponds to dry soil or rocks.

QGIS Functionality to perform PCA on multi-band raster:
- Install the PCA plugin: Plugins > Manage and Install Plugins
- PCA Calculation: menu toolbar Plugins> PCA

2.3.3. Segmentation

Image segmentation is the process used to detect homogeneous objects in remotely sensed images. Image segmentation algorithms look for clusters of similar pixels in an image and group them together based on predetermined criteria into unique and coherent objects (also called segments). The segments contain the mean value (spectral for example) of the pixels that they contain. The shape, size and spectral variance of a segment are useful for classification. There are several tools around, both free and proprietary, which implement different segmentation algorithms. Here, we use a region growing and merging algorithm (GRASS GIS i.segment). All pixels are considered as initial segments. Then, the similarity between any given segment and any of its neighbors is computed using the Euclidean distance. Two segments are merged if the similarity distance is lower than the threshold value given by the user. This process is run iteratively. Finally, segments smaller than the minimum size are merged with their most similar neighbor:

– the segmentation was performed using the three images: PCA1 image, NDVI image and topo-bathymetric DTM.

The following parameters were used:

– a threshold: 0.01;

– minimum number of cells in segment: 200;

– maximum number of iterations: 20.

We use a mask (a vector layer delineates Djebel, Marshes and Lake) in order to constrain segments to the study area. The use of the mask makes it possible to independently segment, on the one hand, the pixels of the lake, the marshes and the Djebel, and on the other hand, the remaining pixels of the study area. In order to build the mask, we rasterized the vector layer that delineates Djebel, Marshes and Lake.

QGIS functionality to create a mask:
- Processing > Toolbox > GDAL/OGR > Conversion > Rasterize

GRASS functionality to realize segmentation:
- Processing > Toolbox > GRASS GIS 7 commands > Imagery > i.segment

Segments are relatively homogeneous with regard to one or more characteristics (Figure 2.8).

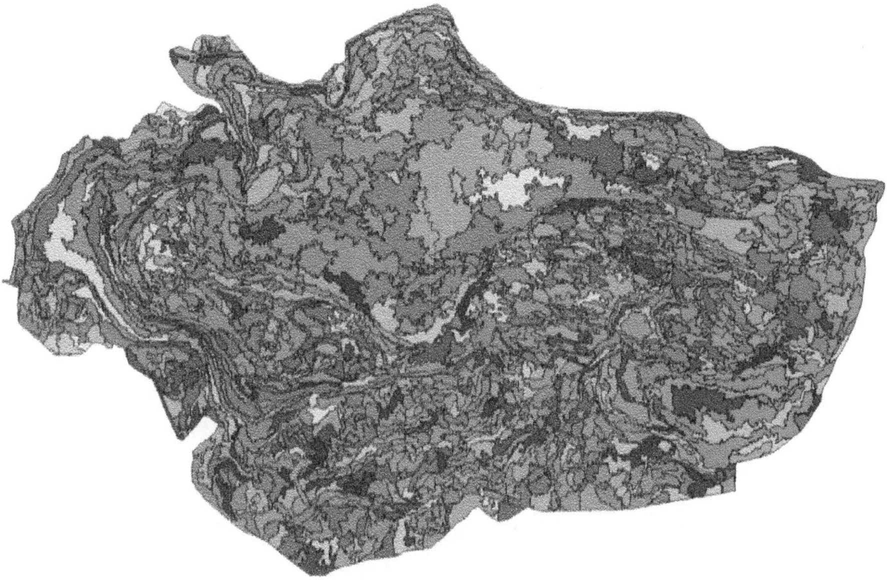

Figure 2.8. *Result of the segmentation process of the ASTER image. The segmented images overlapped with the regions' boundaries. For a color version of the figure, see www.iste.co.uk/baghdadi/qgis4.zip*

In the next step, thematic attributes (average of the mean PCA1, NDVI and topo-bathymetric DTM value) have been computed for each delineated segment. Calculated attributes will be used to determinate classification rules.

QGIS functionality to compute attributes for each segment:
- Install the plugin "zonal statistics": Plugins > Manage and Install Plugins
- To calculate the attribute: Raster > Zonal statistics

2.3.4. Classification

2.3.4.1. *Identity of training sites*

The first step of the classification is to indicate a class name for several segments (Figure 2.9). GPS surveys were carried out to collect samples characterizing the plant communities presented on the study area (about 10 different types of plant). To select segments allowing the construction of the classification rules, the GPS surveys was superposed with the segmentation vector layer (QGIS function "join attributes by location"). Indeed, now joined together, each training segment will have, in addition to the three means (PCA, NDVI and DTM), the name of the class (plant community, body of water, etc.).

QGIS functionality to joint two layers by location:
- Processing > Toolbox > QGIS geoalgorithms > Vector general tools > Join attributes by location

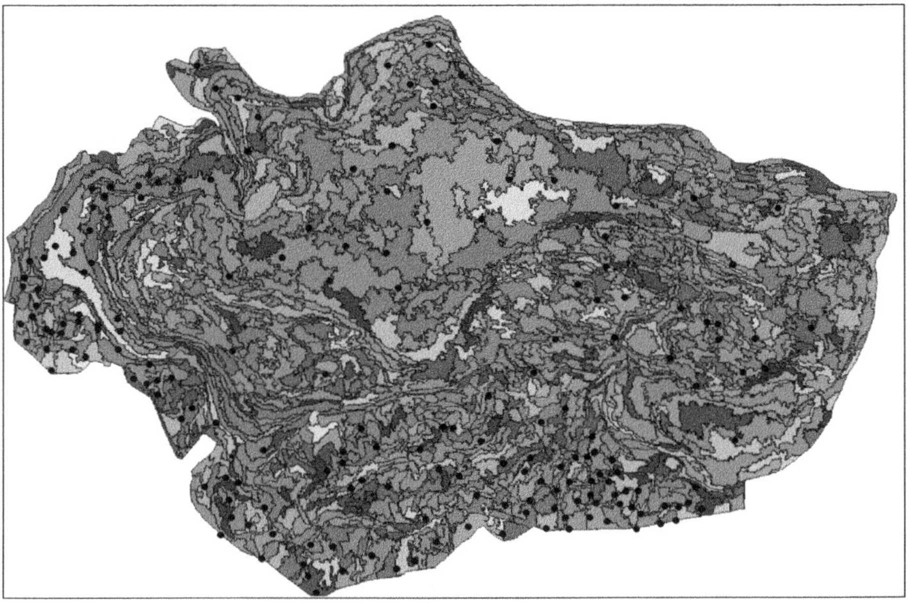

Figure 2.9. *Black points represent the locations of field survey zones (reference objects) that have been used in the study. Polygons are the segments. For a color version of the figure, see www.iste.co.uk/baghdadi/qgis4.zip*

2.3.4.2. *Decision tree classification*

Decision tree classification is a supervised classification method that aims to use rules to classify an image [KIM 16, GIL 08]. In this chapter, classification rules for distinguishing vegetation communities are created automatically by using attributes (PCA1, topo-bathymetric DTM and NDVI value) of training segments. The first rule in the classification tree separates non-vegetated and vegetated objects using the NDVI attribute. Vegetation communities segments have been classified by using the three attributes (a file text containing the classification rules is given with the chapter database).

GRASS Functionality to realize the classification:
- Processing > Toolbox > GRASS GIS 7 commands > Vector > v.reclass

2.3.5. Limitations of the methodology

From the obtained results, some limitations must be highlighted and focus mainly on the following points:

– DTM quality depends on the quality of topographic data used to build the model: spatial distribution, altimetry precision and the protocol of acquisition of altitudinal data;

– vegetation mapping is closely dependent on the image data acquisition;

– classification parameters are fixed by the user;

– the decision tree used for the classification is dependent on the training and validation sites.

2.3.6. Case example of topo-bathymetric transect with the associated vegetation communities

As shown in Figure 2.10, the application of the object-oriented classification in the case of Ichkeul wetland allows the detailed vegetation mapping of marshes, lake and mountain as well as a good hierarchical classification of communities according to their geographic location. The resulting classification distinguishes six major communities, which are as follows: Hordeum communities, Visnaga communities, Bolboschoenus communities, Tamarix communities, Sarconia Frictosa communities and Galactitess communities.

The vegetation communities mapping combined with the topo-bathymetric altitudinal cross-section (Figure 2.11) led us to identify the vegetation distributions according to the altitudinal level. For instance, the marshes are dominated by Spartina patens, Juncus subulateus, Bolboschoenus maritimus and Bolboschoenus litoralis, which require annual inundation, are not salt tolerant and are the main food of graylag geese. And what is more, these are being replaced by more halophytic Visnaga daucoides and Scolymus maculatus. The vegetation of Joumine and Melah marshes ranges from emergent plant associations such as Schoenoplectus lacustris, Typha angustifolia and Phragmites communis through a belt of halophytes, Salicornia arabica, Suaeda maritima and Arthrecnemum fruticosum, to grass pastures with shrubby Tamarix africana. In drier areas and on the well-drained ridges are associations of Hordeum maritimum with Lolium multiflorum and Daucus carota or Nerium oleander and Ziziphus lotus.

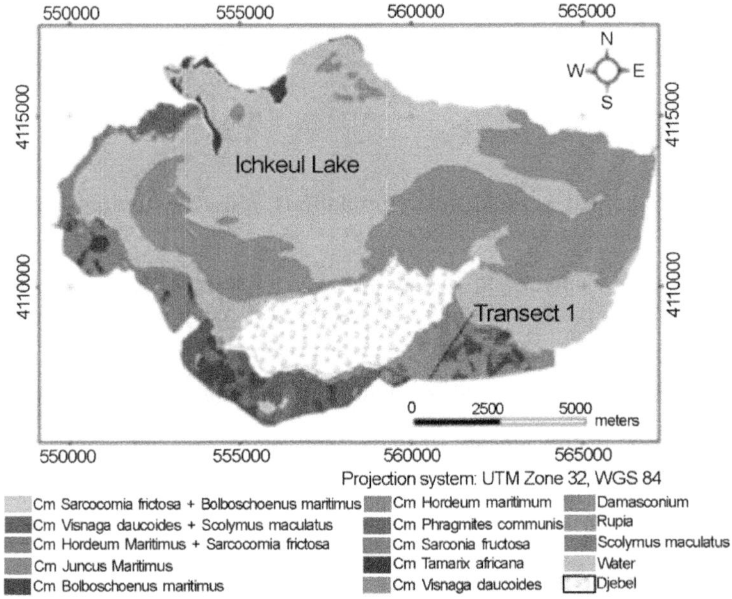

Figure 2.10. *Map of Ichkeul vegetation in June 2007: the three units are simultaneous-mapped using the same data and the same techniques. Vegetation communities are mostly identified in the marshes. For a color version of the figure, see www.iste.co.uk/baghdadi/qgis4.zip*

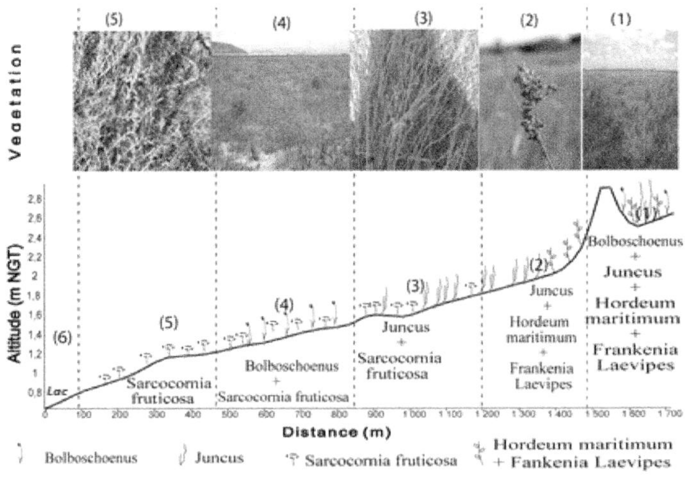

Figure 2.11. *Topographic cross-section of transect1 (see location in Figure 2.10) marsh with their associated vegetation communities. For a color version of the figure, see www.iste.co.uk/baghdadi/qgis4.zip*

2.3.7. Conclusion

The results of this study suggest that ASTER data, in combination with an integrated topo-bathymetric DTM, can be used to map dominant vegetation communities in Ichkeul wetland. Ten communities have been identified and mapped in the marshes; however, in the Djebel Ichkeul as well as in the lake, we only identify general classes (water, Rupia and forest).

2.4. From a practical point of view in QGIS

In this section, we present how we have handled QGIS and GRASS with vegetation communities by using an ASTER image acquired on June 8, 2007.

2.4.1. Software and data

2.4.1.1. *Required software*

In this section, we present how we have handled QGIS and GRASS for vegetation community mapping by using an ASTER image acquired on June 8, 2007.

Methodology	Processing on QGIS
1. Install a plugin in QGIS	In **QGIS:** In the menu toolbar: • Click on "Plugins" → "Manage and install plugins" o Choose the desired extension and then install it o In this chapter, the required extensions are "PCA" and "Zonal Statistics" 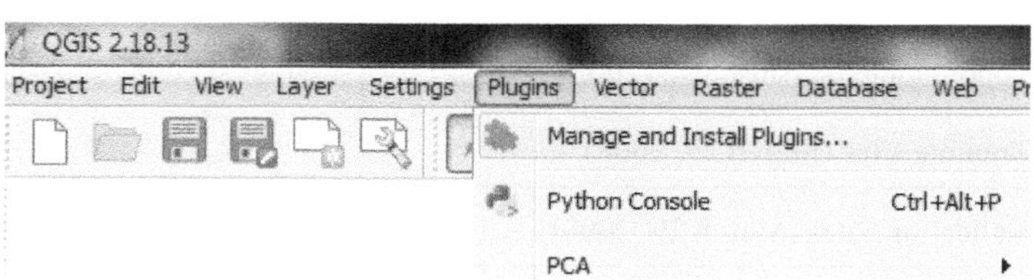

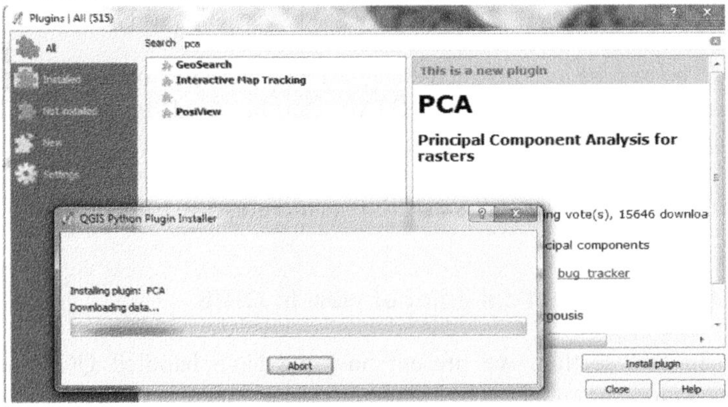

Table 2.4. *Installation procedure of a plugin in QGIS*

2.4.1.2. *Used data*

The data used are of two types: an ASTER satellite image acquired on June 8, 2007, and thematic geographic data in vector format (Table 2.5).

Data	File Name
ASTER image acquired on June 8, 2007[1]	AST_L1B_00306082007101811_20070709230934_11700.hdf AST_L1B_00306082007101811_20070709230934_11700.met
Bathymetric data (2003) for lake	Bathymetry.shp
Topographic data for Djebel and surrounding zone	Contours.shp Points.shp
Topographic marshes data (2003)	MarshesTopography.shp
Training sites (March 29, 2008)	Training_sites.shp
Validation Sites (March 29, 2008)	Validation_sites.shp

Table 2.5. *List of used datasets*

1 ASTER data are available for free dowload in earthexplorer.usgs.gov. For this chapter we downloaded "AST_L1A_ 00306082007101811_20070709230934_11700.hdf".

2.4.2. Computation of the topo-bathymetric DTM

Computing the new topo-bathymetric DTM involves the following steps:

– conversion of data to the same geometry type (point);

– merging of shape file (Points);

– calculation of altitudinal point's density;

– interpolation of merged points to create a DTM.

<table>
<tr><th>Methodology</th><th>Processing on QGIS</th></tr>
<tr><td>1. Convert the contour lines (polyline) to points</td><td>
In QGIS:
<ul><li>Open the vector layer Contours.shp</li></ul>
In the menu toolbar:
<ul><li>Click on "vector" ➔ "Geometry Tools" ➔ "Extract nodes"</li>
<li>Save the file under the name ContoursPts.shp</li></ul>
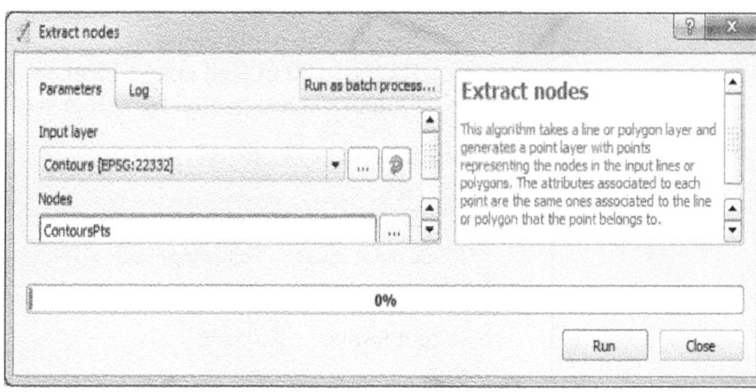

</td></tr>
<tr><td>2. Merge the data</td><td>
In QGIS toolbar:
<ul><li>Open files:
<ul><li>ContoursPts.shp</li><li>Points.shp</li><li>Bathymetry.shp</li><li>MarshesTopography.shp</li></ul></li>
<li>In the menu toolbar:</li>
<li>Click on "Processing" ➔ "Toolbox" ➔ "QGIS geoalgorithms" ➔ "Merge Vector layers":</li>
<li>Click on "Run as batch process" and select all layers</li></ul>
</td></tr>
</table>

<table>
<tr>
<td></td>
<td>

• Save data as: PointsTopobathyDTM.shp

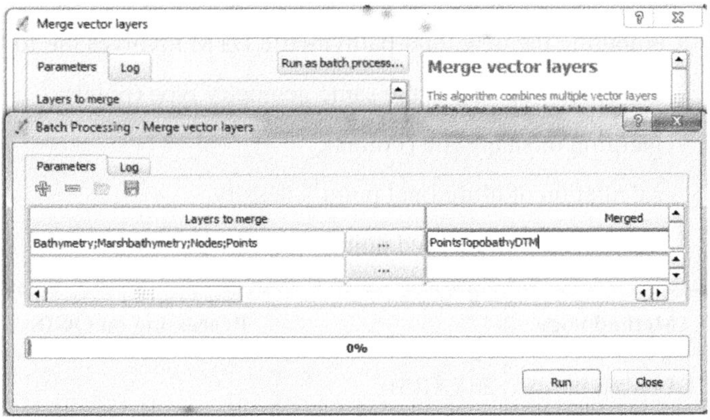

</td>
</tr>
<tr>
<td>3. Build a density map</td>
<td>

In **QGIS:**

• Open the file PointsTopobathDTM.shp

• In the menu toolbar:

o Click on “Raster” ➔ “Heatmap” ➔ “Heatmap”
o Specify the parameters:
o Radius: 100 m/Cell size = 15 m
o Output values: Scaled by kernel size

• Save the file as “Density”

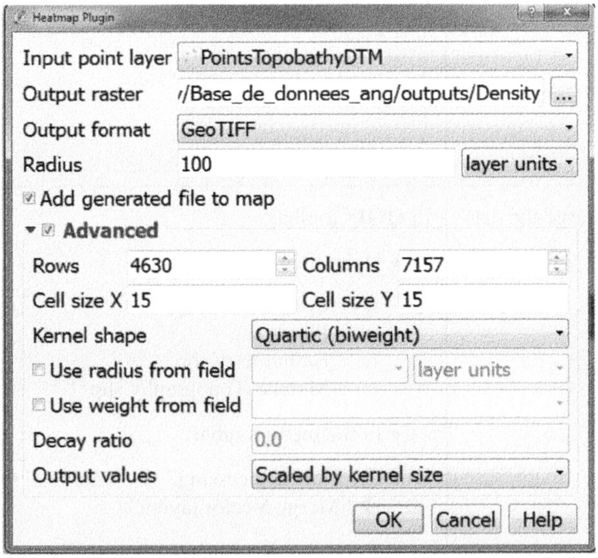

</td>
</tr>
</table>

<table>
<tr>
<td>4. Interpolate the topo-bathymetic DTM (Digital Terrain Model)</td>
<td>In QGIS:
<ul><li>Open the file PointsTopobathyDTM.shp</li></ul>
In the menu toolbar:
<ul>
<li>Click on “Raster” ➔ “Interpolation” ➔ “Interpolation”</li>
<li>Specify the parameters :
<ul>
<li>Cell size X and Cell size Y: 15 m</li>
<li>Interpolation method: TIN</li>
</ul></li>
<li>Save the file as “IchkeulDTMRaster.tif”</li>
</ul>
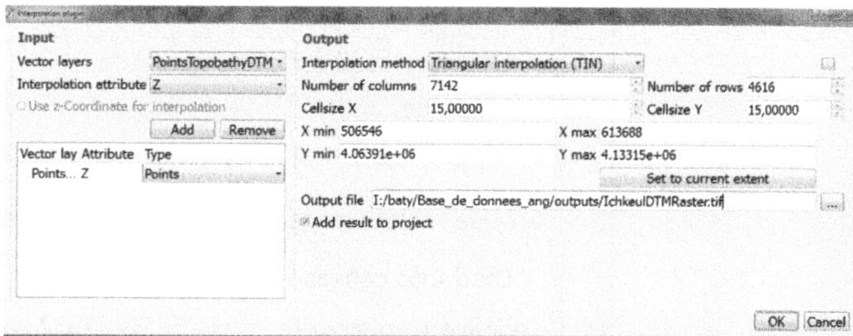
</td>
</tr>
<tr>
<td>5. Reproject the DTM raster layer</td>
<td>The topo-bathymetic DTM (Digital Terrain Model) is Reprojected to UTM 32N:
<ul>
<li>In the menu toolbar, click on “Raster” ➔ “Projections” ➔ “Warp (Reproject)”</li>
<li>In the window that appears, select the raster layer IchkeulDTMRaster.tif in “Input file”</li>
<li>In “output file”, enter IchkeulDTMRaster_utm32n.tif</li>
<li>For “Target SRS”, choose “WGS84/UTM zone 32N (EPSG: 32632)”</li>
<li>Finally, click on “OK”.</li>
</ul></td>
</tr>
</table>

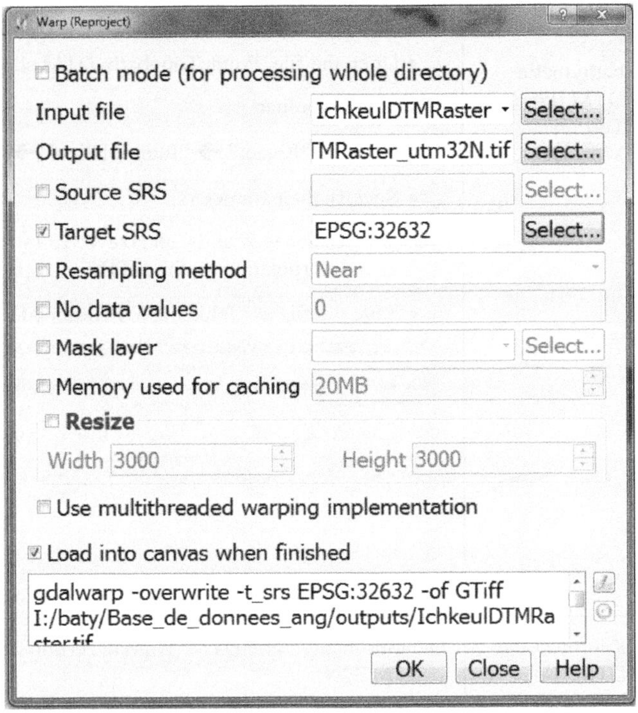

Table 2.6. *Steps for the construction of the topo-bathymetric DTM*

2.4.3. Image preprocessing

2.4.3.1. *ASTER image band extraction and atmospheric correction*

Methodology	Processing on QGIS
1. Extraction of the bands from the HDF image, and calculation of reflectance at the Top of the Atmosphere (TOA)	A python script is developed to calculate the reflectance at the top of the atmosphere for the first ten bands of the ASTER image (processing level L1B, HDF format). ➤ Add the developed script in **QGIS**: • In the menu toolbar, click on “Processing”➔ “Toolbox”. • In the window that appears, click on “Scripts” ➔ “Tools” ➔ “Add script from file” and select the script “ASTER_LB_rTOA.py”. This script is available with the chapter database

<table>
<tr>
<td></td>
<td>

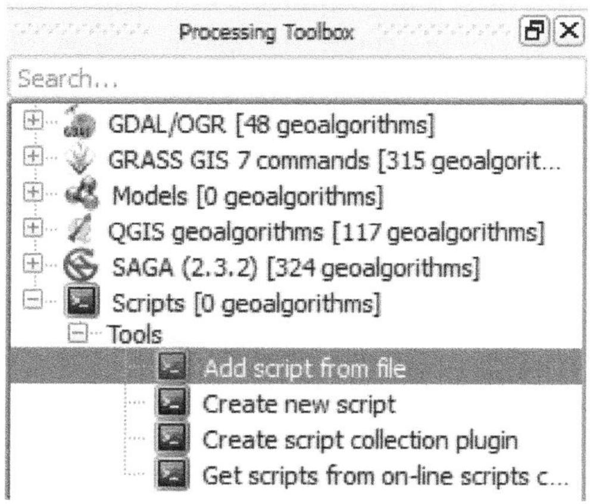

➢ Run the script from the QGIS interface
• In the menu toolbar, click on “Processing”➔ “Toolbox”.
• In the window that appears, click on “User scripts” to display the script added in the previous step “ASTER_LB_rTOA.py”
• Right click on the script “ASTER_LB_rTOA.py” ➔ “Run”
• In the window that appears, enter the path of the ASTER image name (hdf format), and the sun elevation angle. The sun elevation angle is provided in the xml file containing the metadata of the ASTER image

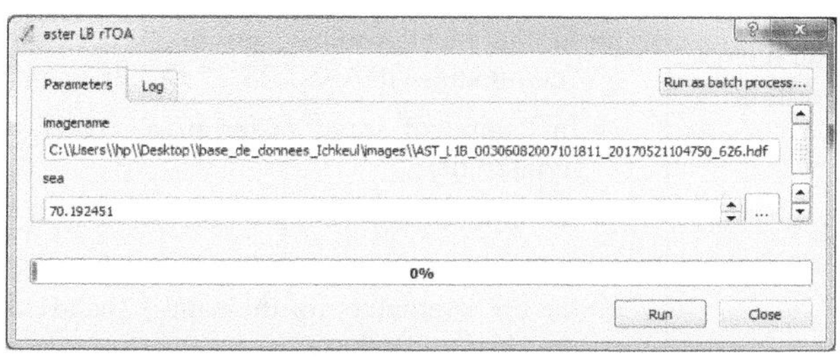


```
<MDI key="SIZEMBDATAGRANULE">119.233</MDI>
<MDI key="SOLARDIRECTION">133.261862, 70.192451</MDI>
<MDI key="SOURCEDATAPRODUCT">ASTL1A 0706081018110706120368,
<MDI key="SOUTHBOUNDINGCOORDINATE">36.936769103738</MDI>
<MDI key="SPATIALRESOLUTION">15, 30, 90</MDI>
```

• Finally, click on “Run”. ASTER bands are automatically saved in a folder called “rtoa”. This folder is created by the script in the same directory as the ASTER image in HDF format.

</td>
</tr>
<tr>
<td>2. Reprojection of bands obtained in the previous step</td>
<td>

➢ To reproject bands (Top of Atmospheric reflectance) :
• In the menu toolbar, click on “Raster” ➔ “Projections” ➔ “Warp (Reproject)”
• In the window that appears, select band_1.tif in “input file”.

</td>
</tr>
</table>

	• In "output file" , enter "band_1_utm32n.tif" • Activate the option "Target SRS" , and choose the project "WGS84/UTM zone 32N (EPSG: 32632)" 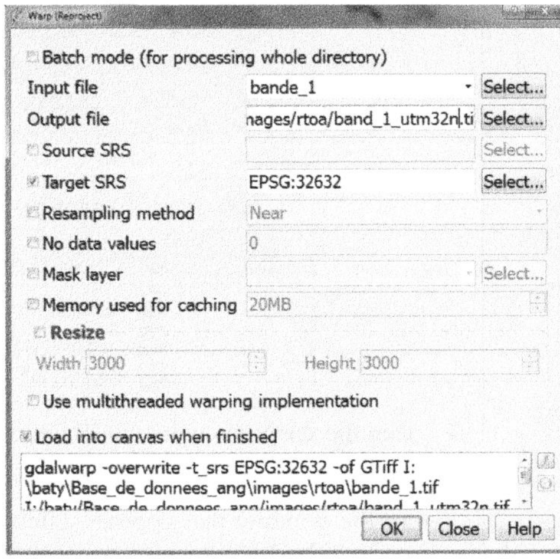 • Click on "OK". • Repeat the above steps for bands 2 (band_2_utm32n.tif), 3N (band_3n_utm32n.tif), 4 (band_4_utm32n.tif) and 5 (band_5_utm32n.tif).
3. Atmospheric correction using the 6S algorithm	➤ To correct for atmospheric effects: • In the menu toolbar, go to "Processing" ➔ "Toolbox" ➔ "commands GRASS GIS 7" ➔ "Imagery" ➔ "i.atcorr". • In "name of input raster map", enter the first band (band_1_utm32n.tif) • In "input imagery range" enter 0 in "Min" and 1 in "Max". • In "name of input text file", enter the file (band1.txt) that contains the 6S parameters for the band 1 (band1.txt). This file is given with the chapter database. • In "Rescale output raster map" enter 0 in "Min" and 1 in "Max". • In "Atmospheric correction" enter the name of the output image: band_1_utm_32n_sr.tif • Click on "OK" • Repeat the above steps for bands 2, 3N, 4 and 5. For each band, a text file containing the parameters of the 6S model is provided with the chapter database.

<table>
<tr><td></td><td>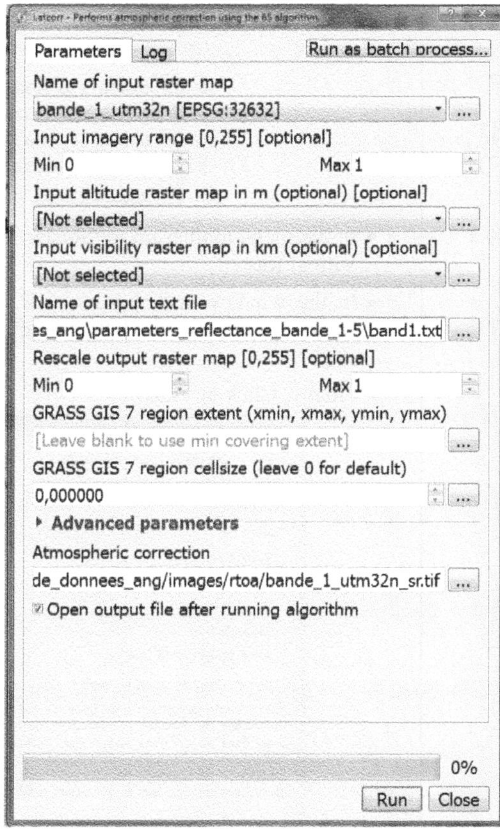
</td></tr>
<tr><td>4. Layer stacking of bands corrected from atmospheric effects</td><td>To facilitate further utilization of ASTER bands corrected for atmospheric effects, these bands were stacked and then clipped according to the extent of the study area.

➢ Layer stacking of bands 1, 2, 3N, 4 and 5 corrected from atmospheric effects:

• Import the five bands into QGIS: band_1_utm_32n_sr.tif, band_2_utm_32n_sr.tif, band_3n_utm_32n_sr.tif, band_4_utm_32n_sr.tif and band_5_utm_32n_sr.tif

• In the “layers panel” section of the QGIS interface, order the mono-band images from top to bottom as below:
▪ band_1_utm_32n_sr.tif,
▪ band_2_utm_32n_sr.tif,
▪ band_3n_utm_32n_sr.tif,
▪ band_4_utm_32n_sr.tif,
▪ band_5_utm_32n_sr.tif</td></tr>
</table>

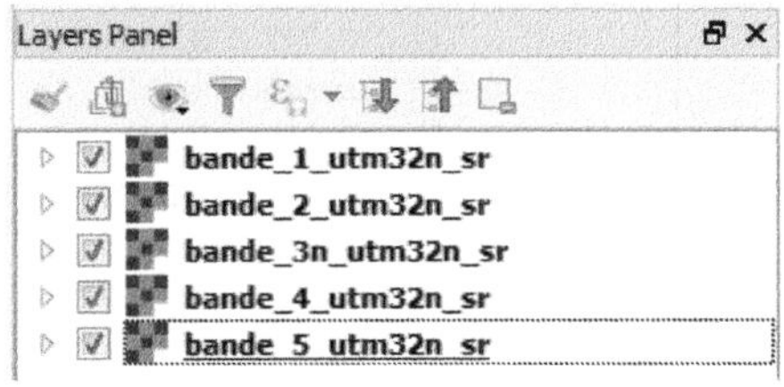

- In the menu toolbar go to "Raster" ➔ "Miscellaneous" ➔ "Build Virtual Raster"
- In the window that appears, tick the options "Use visible raster layers for inputs", and "separate". Then, name the output file: stack.vrt
- Finally, Click on "OK"

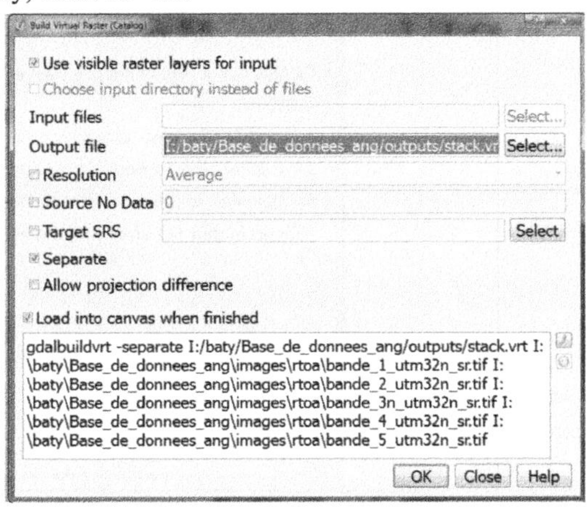

➢ To clip the stacked image stack.vrt according to the extent of the study site:

- Import the stack.vrt raster layer
- In the menu toolbar, click on "Raster" ➔ "Extraction" ➔"Clipper"
- In the window that appears, enter the extent coordinates:

	x (East)	y (North)
North-West Corner	547910	4118570
South-East Corner	568790	4102520

- Name the output file stack_Ichkeul.tif
- Finally, click on "OK"

	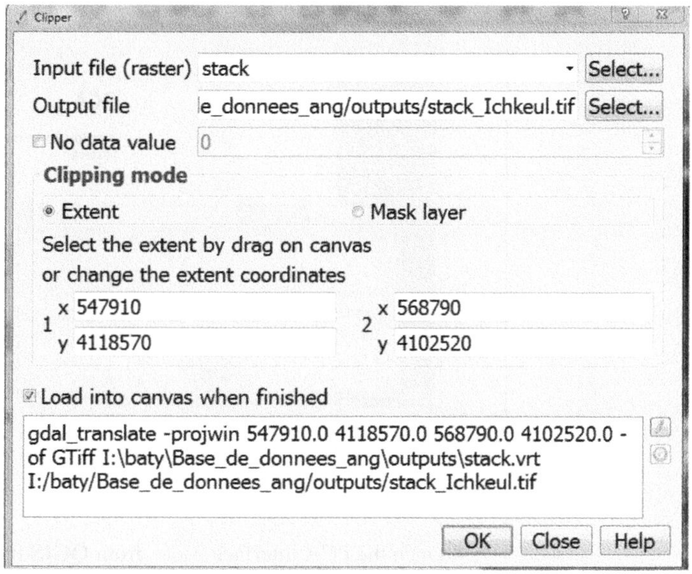

Table 2.7. *Preprocessing steps of the ASTER image: extraction, projection, atmospheric correction, stacking of bands (1, 2, 3N, 4 and 5) and clip according to the study area. For a color version of the table, see http://www.iste.co.uk/baghdadi/qgis4.zip*

2.4.3.2. *Principal component analysis*

Methodology	Processing on QGIS
1. Computation of principal component analysis (PCA)	➤ Install the PCA plugin: • In the menu toolbar, click on “Plugins” ➔ “Manage and Install Plugins” • In the window that appears, write PCA and install the extension • Click on “Install plugin”

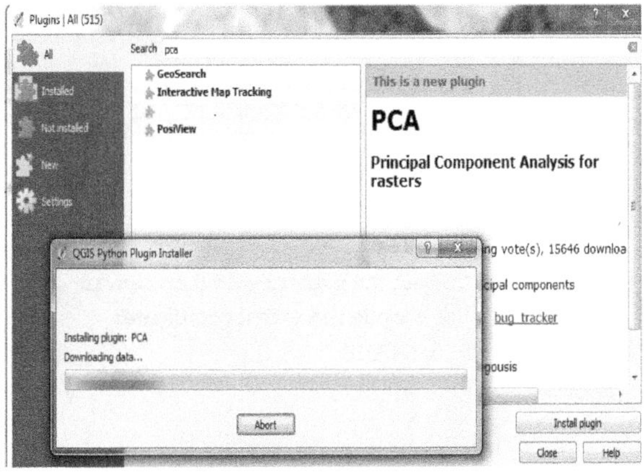

- Finally, click on “Close”

➢ To calculate the PCA image:

- Open the PCA interface from **QGIS** interface
- In the window that appears, enter the image stack_Ichkeul.tif in “Input raster file”
- In “Number of output principal components”, choose 1 to obtain only the first component
- In “Output raster file”, enter PCA_1c_Ichkeul.tif
- Click on “OK”

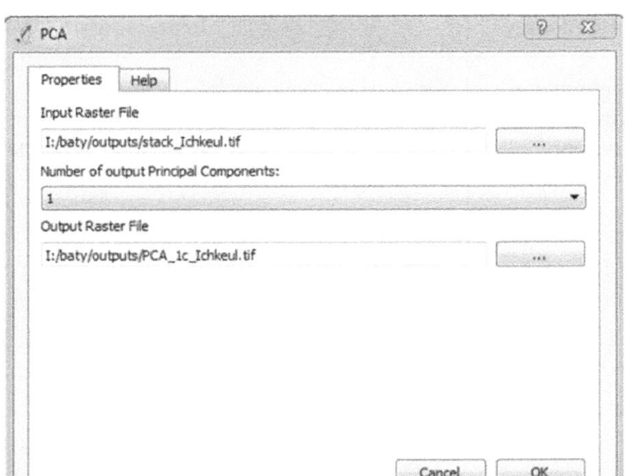

Table 2.8. *Computation of ACP image*

2.4.3.3. *Calculation of the Normalized Vegetation Index (NDVI)*

Methodology	Processing on QGIS
1. Calculation of the Normalized Vegetation Index (NDVI)	➢ To calculate the NDVI: • Import the raster file stack_Ichkeul.tif in **QGIS** • In the menu toolbar, click on "Raster" ➔ "Raster Calculator" • In "Raster calculation expression", write: ("stack_Ichkeul@3"-"stack_Ichkeul@2")/ ("stack_Ichkeul@3"+"stack_Ichkeul@2") • In "Output layer" give the name ndvi_Ichkeul.tif • Finally, click on "OK" 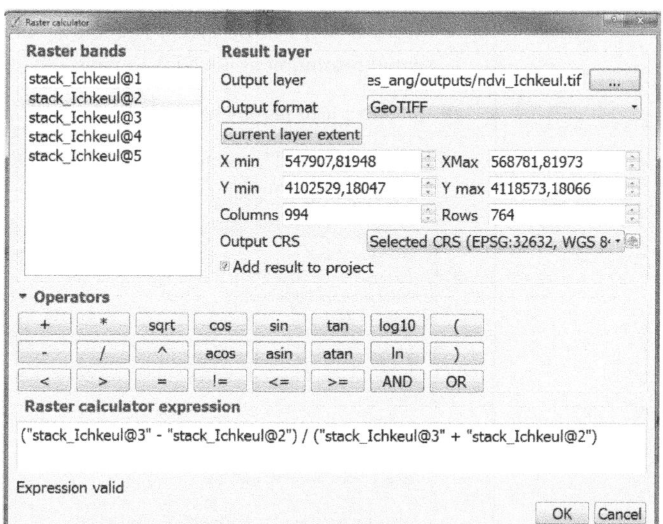

Table 2.9. *Calculation of the Normalized Vegetation Index (NDVI)*

2.4.4. Segmentation

The segmentation is applied to the image formed by the first component of the PCA, the computed topo-bathymetric numerical model (raster format) and the NDVI image. The vector file delimiting the three Ichkeul units (lake, marshes and Djebel) is also introduced into the segmentation process as a thematic layer to force segmentation. The thematic layer allows us to independently segment the pixels corresponding to the lake zone (lake, marshes, and Djebel) and the pixels of the rest of the study area.

<table>
<tr><th>Methodology</th><th>Processing on QGIS</th></tr>
<tr><td>1. Segmentation</td><td>

➢ To build the mask image:

- Open the vector layer MarshLackeDjebel.shp. This layer is given with the chapter database
- In the menu toolbar, go to "Processing" ➔ "Toolbox" ➔ "GDAL/OGR" ➔ "Conversion" ➔"Rasterize"
- In "Input layer", enter the vector layer MarshLackeDjebel.shp
- In "Attribute field", choose the Id field
- In "Horizontal" and in "Vertical", enter 15 to obtain an output image with 15 m of spatial resolution, which corresponds to the spatial resolution of the NDVI image.
- In "Advanced parameters" ➔ "raster type", choose Byte
- In "Rasterized", enter the name of the output image LackeMarshesDjebel_raster.tif

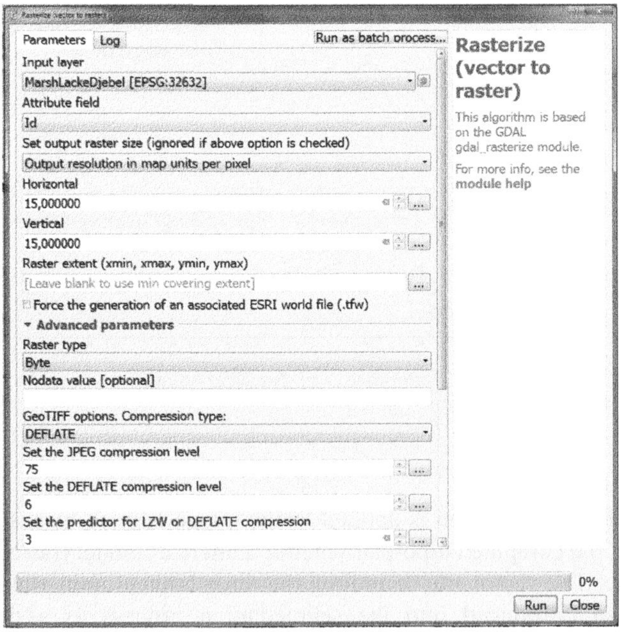

</td></tr>
</table>

<table>
<tr>
<td></td>
<td>

➢ To segment:

- Open the following raster files: ndvi_Ichkeul.tif, PCA_1c_Ichkeul.tif, IchkeulDTMRaster_utm32N.tif, and LackeMarshesDjebel_raster.tif
- In the menu toolbar, click on "Processing" ➔ "Toolbox" ➔ "Commandes GRASS GIS 7" ➔ "Imagery" ➔ "i.segment"
- In "Input Rasters", choose: ndvi_Ichkeul.tif, PCA_1c_Ichkeul.tif, IchkeulDTMRaster_utm32N.tif
- In "Difference threshold between 0 and 1", enter 0.01
- In "Minimum number of cells in a segment", enter 200
- In "Name of input raster bounding/constraining raster map", select the mask image LackeMarshesDjebel_raster.tif
- In "Segmented raster", enter the name of the output image: segmentation.tif
- Click on "Run"

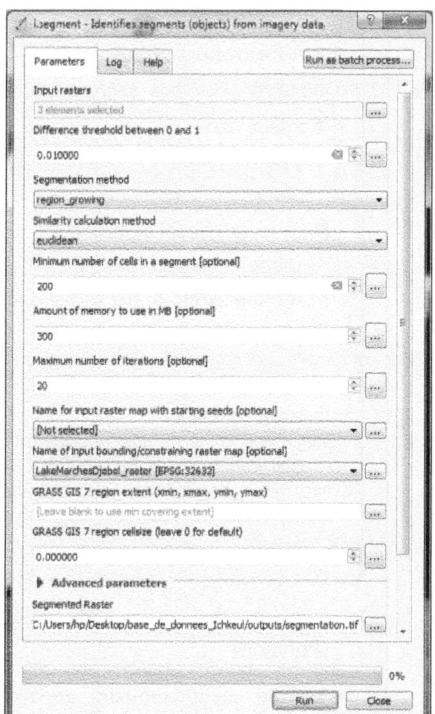

</td>
</tr>
</table>

➤ To keep only the segmentation performed on the lake zone, the segmentation image is clipped:

- Open the raster layer segmentation.tif and the vector MarshLackeDjebel.shp
- In the menu toolbar, click on "Raster" ➔ "Extraction" ➔"Clipper"
- In the window that appears, select the segmented image (segmentation.tif) in "Input file (raster)"
- In "output file", give the output image the following name: segmentation_Ichkeul.tif
- Tick the option "Mask layer", and enter the name of the vector layer MarashesLacDjebel.shp
- Tick the option "Crop the extent of the target dataset to the extent of the cutline"
- Finally, click on "OK"

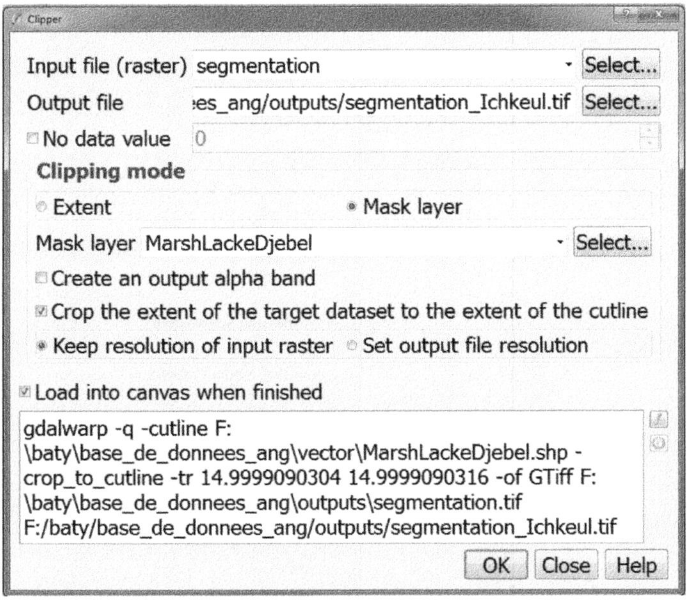

➤ Convert the raster file segmentation_Ichkeul.tif to vector (Polygon):

- In the menu toolbar, go to "Raster" ➔ "Conversion" ➔ "Polygonize".

	• In the window that appears, select the segmentation image (segmentation_Ichkeul.tif) in "Input file (raster)" • In "Output file for polygons (shapefile)", enter the name: Ichkeul_Seg.shp • Finally, click on "OK" 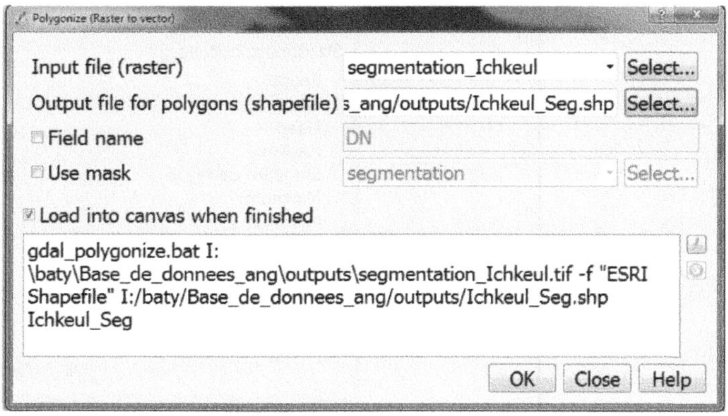
2. Compute attributes for each segmentation polygon	➤ Determine for each segment the mean of NDVI: • Open the raster layers: ndvi_Ichkeul.tif, PCA_1c_Ichkeul.tif and IchkeulDTMRaster_utm32N.tif • In the menu toolbar, click on "Raster" ➔ "zonal statistics" ➔ "zonal statistics" • In the window that appears, select the raster layer ndvi_Ichkeul.tif in "Raster layer" • In "polygon layer containing the zones", select the vector Ichkeul_Seg.shp • In "output column prefix", write "NDVI_" • In "statistics to calculate", tick the option "mean" • Click on "OK". Once the processing is complete, the values of the mean are saved in the attribute table.

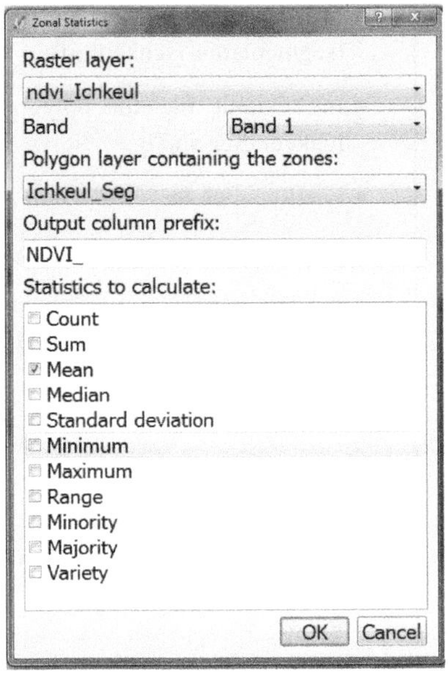

- Repeat the previous process for PCA_1c_Ichkeul.tif "PCA_" and IchkeulDTMRaster_utm32N.tif "DTM_"
- Open the attribute table to check results

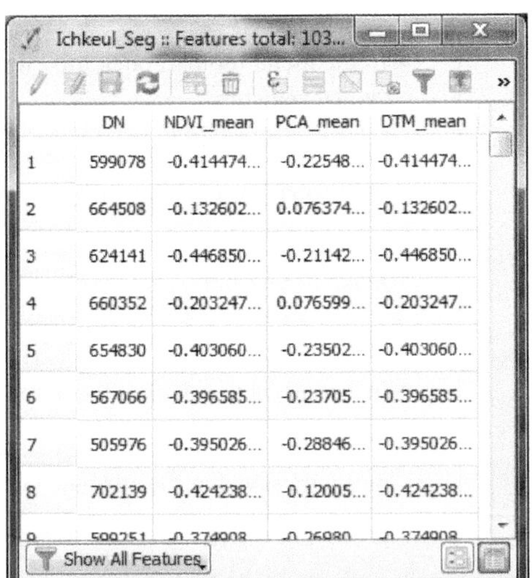

Table 2.10. *Segmentation and calculation of the attributes for each segmentation polygon*

2.4.5. Classification

2.4.5.1. *Identification of training sites*

This step consists of identifying the training zones through their locations and their attributes, and proposing a classification of the whole image. It consists of identifying the objects (segmentation polygons) that contain zones determined *in situ*.

Methodology	Processing on QGIS
1. Determination of training polygons and assignment of class names	In **QGIS**: • Open Training_sites.shp and Ichkeul_Seg.shp • In the menu toolbar, click on "Processing"➔ "Toolbox" ➔ "QGIS Geoalgorithms" ➔ "Vector general tools" ➔ "Join attributes by location" • In "Target vector Layer", select the vector layer Ichkeul_Seg.shp • In "Join vector layer", select the vector Layer Training_sites.shp • In "Geometric Predicate", tick "Intersects" • In "Joined Table", tick the option "Only keep matching records" • In "Joined layer", enter the name of output vector layer: Training_SitesPolygon.shp 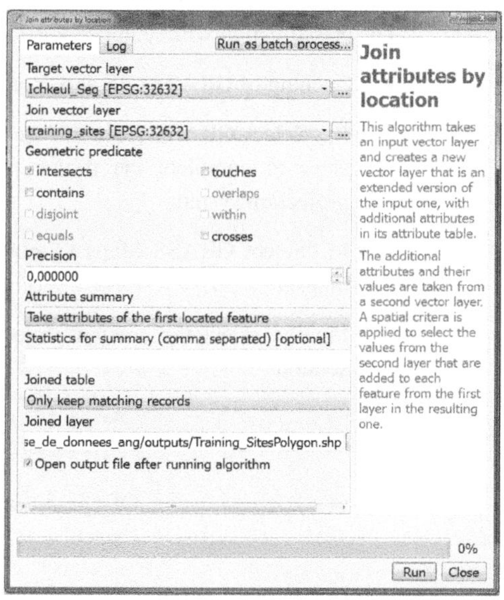

	• To check the results, open the attribute table of the vector layer Training_SitesPolygone.shp:

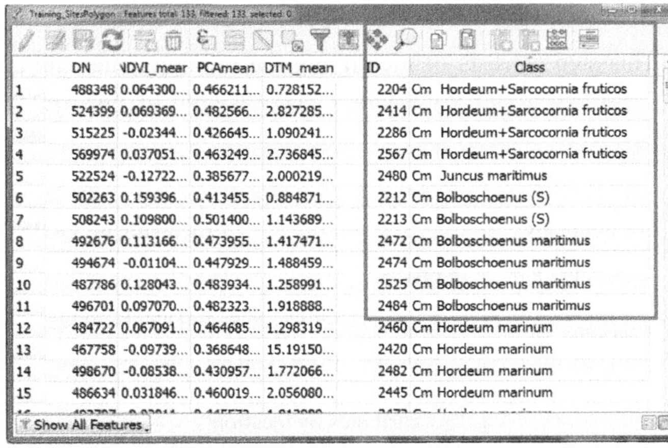

Training_SitesPolygon :: Features total: 133, filtered: 133, selected: 0

	DN	NDVI_mean	PCAmean	DTM_mean	ID	Class
1	488348	0.064300...	0.466211...	0.728152...	2204	Cm Hordeum+Sarcocornia fruticos
2	524399	0.069369...	0.482566...	2.827285...	2414	Cm Hordeum+Sarcocornia fruticos
3	515225	-0.02344...	0.426645...	1.090241...	2286	Cm Hordeum+Sarcocornia fruticos
4	569974	0.037051...	0.463249...	2.736845...	2567	Cm Hordeum+Sarcocornia fruticos
5	522524	-0.12722...	0.385677...	2.000219...	2480	Cm Juncus maritimus
6	502263	0.159396...	0.413455...	0.884871...	2212	Cm Bolboschoenus (S)
7	508243	0.109800...	0.501400...	1.143689...	2213	Cm Bolboschoenus (S)
8	492676	0.113166...	0.473955...	1.417471...	2472	Cm Bolboschoenus maritimus
9	494674	-0.01104...	0.447929...	1.488459...	2474	Cm Bolboschoenus maritimus
10	487786	0.128043...	0.483934...	1.258991...	2525	Cm Bolboschoenus maritimus
11	496701	0.097070...	0.482323...	1.918888...	2484	Cm Bolboschoenus maritimus
12	484722	0.067091...	0.464685...	1.298319...	2460	Cm Hordeum marinum
13	467758	-0.09739...	0.368648...	15.19150...	2420	Cm Hordeum marinum
14	498670	-0.08538...	0.430957...	1.772066...	2482	Cm Hordeum marinum
15	486634	0.031846...	0.460019...	2.056080...	2445	Cm Hordeum marinum

Show All Features

Table 2.11. *Identification of training zones (by their attributes) on the segmented image*

2.4.5.2. *Decision tree classification*

Methodology	Processing on GRASS GIS
1. Decision tree classification	The classification will be done by using GRASS GIS ➢ Opening of GRASS GIS and creation of a project: • In "Select GRASS Location", click on "New", and follow the steps to complete the project setup. Choose the UTM 32N projection system. • In "Select GRASS Mapset", click on "New" to create a new mapset • Finally, select the created project and click on "Start GRASS session"

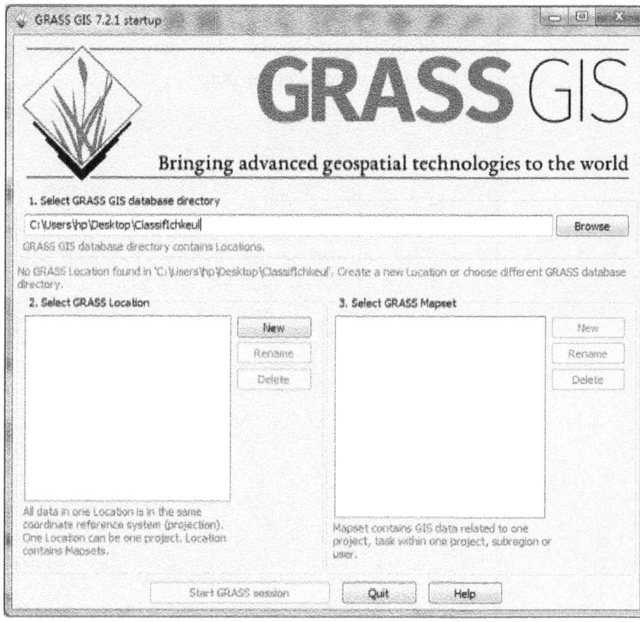

➢ Dataset import and classification:

- Go to "File" ➔ "Import vector data" ➔ "v.import"
- In the window that appears, click on "Browse" and select the vector layer Ichkeul_seg.shp

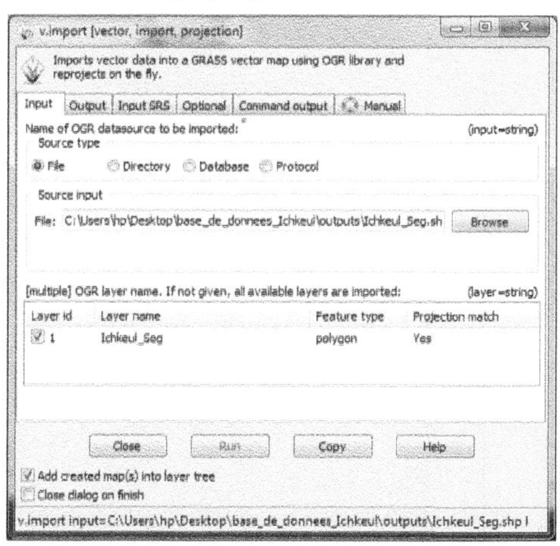

- In the menu toolbar, click on "vector" ➔ "Manage Categories" ➔ "Reclassify (v.reclass)"

- In “Required”, select the vector file Ichkeul_seg.shp in “Name of the Input Vector Map” and write the classification in "Name of the output vector map"

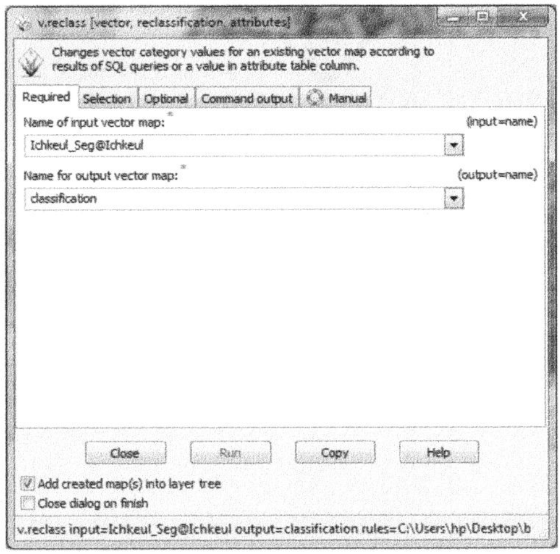

- In “Optional”, specify the path to the text file containing the rules of classification (RulesClasses.txt), in “Full path to the reclass rule file”. The text file RulesClasses.txt is provided with the chapter database.

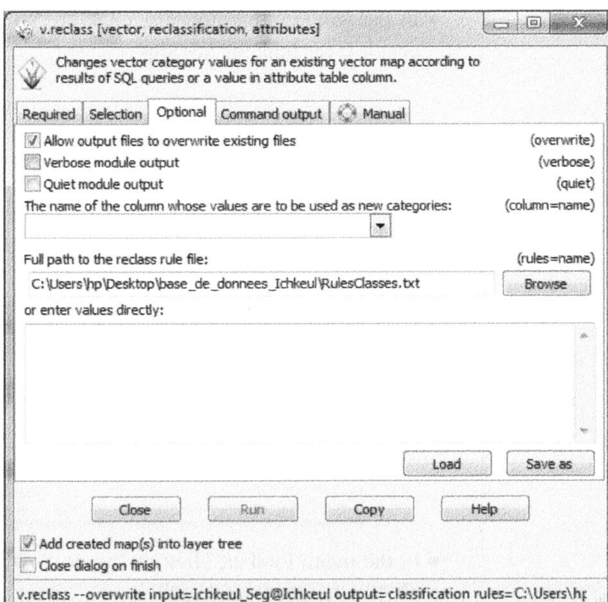

	• Finally, click on "Run" ➤ Export the classification result • In the menu toolbar, click on "File" ➔ "Export vector map" ➔ "Common export format" • Name the output vector layer as follows: classification.shp.

Table 2.12. *Decision tree classification of the segmented image*

2.4.5.3. *Topographical cross-section of the vegetation*

In this part, an example of topographical cross-section of the vegetation is shown.

Methodology	Processing on QGIS
1. Topographical cross-section of transect building	In the map display window of **GRASS GIS**, click on "Profile surface map" and then "Profile surface map" and choose the raster IchkeulDTMRaster.tif 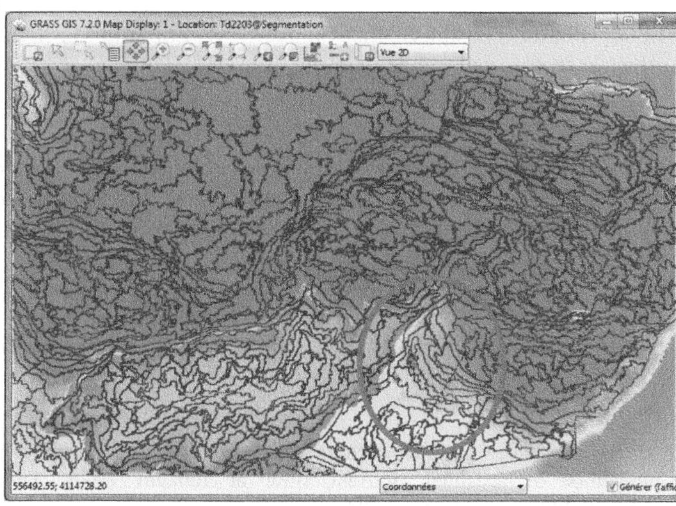 A topographic cross section transect is shown from Joumine marsh (see figure before)

	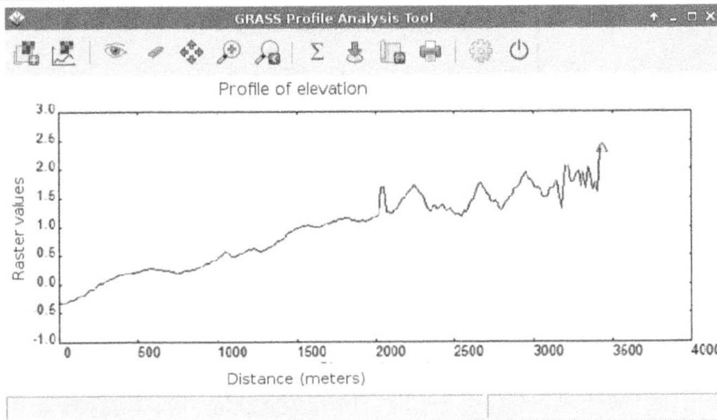

Table 2.13. *Procedure for plotting a topographic profile. For a color version of the table, see http://www.iste.co.uk/baghdadi/qgis4.zip*

2.5. Bibliography

[GIL 08] GILMORE M.S., WILSON E.H., Barrett N. *et al.*, "Integrating multi-temporal spectral and structural information to map wetland vegetation in a lower Connecticut River tidal marsh", *Remote Sensing of Environment*, vol. 112, no. 11, pp. 4048–4060, 2008.

[GHR 06] GHRABI GAMMAR Z., LILI-CHABAANE Z., ZOUAGHI M., "Evolution de la couverture végétale du parc national de l'Ichkeul (Tunisie)", *Revue Ecologique Terre et Vie*, vol. 61, no. 4, pp. 317–326, 2006.

[KAS 08] KASSOUK Z., LILI CHABAANE Z., DEFFONTAINES B. *et al.*, "Exploitation du MNT topo-bathymétrique intégré et des images satellitaires multi sources pour le suivi de la végétation des marais de l'Ichkeul (Nord Tunisie)", *Les XIèmes Journées Scientifiques du Réseau Télédétection de l'AUF*, Madagascar, November 3–7, 2008.

[KIM 16] KIM K., "A hybrid classification algorithm by subspace partitioning through semi-supervised decision tree", *Pattern Recognition*, vol. 60, pp. 157–163, 2016.

[ROU 74] ROUSE J.W., JR. HAAS R.H., SCHELL J.A. *et al.*, "Monitoring Vegetation Systems in the Great Plains with Erts", *Third Earth Resources Technology Satellite-1,* vol. 1, p. 309, Washington, DC, 1974.

[TAM 94] TAMISIER A., BOUDOURESQUE C.F., "Aquatic bird populations as possible indicators of seasonal nutrient flow at Ichkeul Lake, Tunisia", *Hydrobiologia*, vol. 279/280, pp. 149–156, 1994.

[VEN 02] VENABLES W.N., RIPLEY B.D., *Modern Applied Statistics with S*, Springer, New York, 2002.

3

Reservoir Hydrological Monitoring by Satellite Image Analysis

3.1. Context and scientific issue

3.1.1. *Scientific issue*

The objective of this chapter is to monitor the surface evolution of an artificial reservoir during two hydrological years. This monitoring will make it possible to estimate the dynamics and amplitude of the hydrological seasonal variations of this reservoir. This evolution will be compared with the vegetation status in the surrounding areas of the reservoir. The comparison will help to identify the level of water uses for crop irrigation according to the following hypotheses:

– if the reservoir surface is at its maximum at the same time as the vegetation is also at its maximum, and the reservoir at its minimum when the vegetation is inactive, then the water is not used for irrigation;

– conversely, if the reservoir's level is decreasing while the vegetation index grows, then we can suppose that the water is used for irrigation;

– finally, if no correlation or anti-correlation is found, several others factors must be considered.

3.1.2. *Physical and human context*

The study is focused on the Singur reservoir, built in the late 1990s along the Manjira River. Located at 65 km N-W of the Telangana state capital, Hyderabad (India). The maximum capacity of the reservoir is 847 Mm^3. The Singur Lake is

---

Chapter written by Paul PASSY and Adrien SELLES.

mainly used for supplying drinking water to Hyderabad. Since the reservoir's implementation, the population of Hyderabad has doubled to 8 million inhabitants (2011). Nevertheless, part of the stored water is used for irrigation downstream of the dam, mainly for paddy fields.

3.1.3. *The importance of water resources in Central India*

The Singur reservoir is located in a semi-arid region, on the Deccan plateau, at an altitude of around 540 m. Average annual precipitation ranges from 700 to 800 mm. Most of the rain occurs during the monsoon period, from June to October. Since the creation of Hyderabad in 1591, natural and artificial tanks and lakes have been used to supply water for domestic and irrigation purposes. Since the Green Revolution in the 1960s, agriculture has led to increased pressure on both surface water and groundwater resources [CEL 10]. Historically supplied by surface water from the reservoirs connected by complex canal systems, Hyderabad supplied 13% of the rice production in India [ADU 11]. Highly dependent of the monsoon regime and its large variation in intensity, Hyderabad and the rural areas around are facing an increased need for water for domestic use and agricultural practices.

3.2. Methods and data set

3.2.1. *Methods*

In this chapter, we use QGIS software, version 2.18.3. The first step is to establish a chain of treatments to extract the water surface of the reservoir and assess its area. Then, the vegetation status is analyzed via a calculation of Vegetation index. Eventually, we combine both operations to obtain a spatio-temporal view of these phenomena. The chain of treatments is established via QGIS interface, allowing for a semi-automatic process during the different steps. To limit the amount of information treated here, four Landsat 8 satellite images will be detailed in this chapter corresponding to four dates: two images at the end of the rainy season and two at the end of the dry season. The chain of treatments contains the following steps:

1) Calculation of the Automated Water Extraction Index (AWEI).

2) Building a binary raster from the AWEI raster to obtain the water bodies identified by pixels with a 1 value and 0 for the others.

3) Vectorization of the former binary raster.

4) Selection of polygons representing the water bodies.

5) Calculation of the areas of each water body in km^2.

6) Selection of the water surface of the Singur Reservoir.

7) Calculation of the Soil-Adjusted Vegetation Index (SAVI) to characterize the vegetation status.

8) Building a spatial mask on the land surfaces to only take the water bodies during the calculation of the average value of SAVI.

9) Application of this mask on the SAVI raster.

Calculation of the mean SAVI for the land area of the study zone.

3.2.2. *Data set*

For this chapter we use Landsat 8 images distributed free of charge by NASA. For details on this sensor, the reader can refer to the section of this book focused on using the SAGA GIS modules [PAS 18]. We use images already pre-treated in surface reflectance, which avoids the different steps of radiometric and atmospheric corrections. If the reader does not wish to download the data themselves, the images already compiled and cropped on the study area are provided with this chapter. In this case, it is possible to go directly to section 3.3 of this chapter.

These surface reflectances are available on the EarthExplorer portal, maintained by the USGS [PAS 18]. The study area is at the intersection of "path 144" and "row 48". We focus on the period from November 2014 to May 2014. In the "Data Sets" tab of the portal we select "Landsat Archive > Collection 1 Higher-Level (On-Demand) > L8 OLI / TIRS C1 Higher Level". In the "Additional Criteria" tab we request images with less than 20% cloud cover ("Land Cloud Cover less than 20%").

With all these criteria, 22 scenes are found. It is appropriate to use the 22 scenes, but for this chapter we will only use the 4 on the following dates:

– November 26th 2014

– May 21st 2015

– November 13th 2015

– May 23rd 2016

For each selected scene, we click on the "Order Scene" cart icon that changes color after selection. Once the 4 scenes have been selected, we click on "Submit Standing Request". On the new page we click on "Item Basket (4)" and on "Proceed To Checkout". As we request images with a high pre-processing level, they are not immediately available. The treatment takes anywhere from a few minutes to a few hours. An initial confirmation email is sent, followed by a second message that the images are being processed, and finally a final email alerts us when the images are ready. This last e-mail contains a link leading to a page where it is possible to download the different scenes. Once this download is complete, it is necessary to decompress the archives corresponding to each of the scenes.

At this stage we have 4 directories, corresponding to our 4 scenes, named:

– LC08144048201411260lT1-XXX

– LC081440482015052101T1-XXX

– LC081440482015111301T1-XXX

– LC081440482016052301T1-XXX

In each of the directories, we find the first seven Landsat 8 bands named xxx_band1.tif and xxx_band7.tif (deep blue, blue, green, red, near infrared, medium infrared 1 and medium infrared 2 respectively). We also have a metadata file "xxx_MTL.txt". This is a text file containing a lot of information, including the date of the scene acquisition at the "DATE_ACQUIRED" line.

If we open one of the downloaded bands in QGIS, we realize that pixel values range from 0 to several thousand. The reflectances are by definition numbers between 0 (no reflectance) and 1 (reflectance of 100%). Here, these values have been multiplied by 10,000 in order to avoid manipulating decimal numbers.

The projection system used is the WGS84/UTM system 44N (EPSG 32644). This system has the advantage of being expressed in meters.

3.2.3. *Data set preparation*

In order to reduce processing times and the manipulated volume of data, we will cut our images according to the boundaries of our study area. These boundaries are in the shapefile "extent_singur_32644.shp". Let us start by importing this shapefile into QGIS as well as the 7 Landsat spectral bands of the first date (26/11/2014) stored under the name "LC081440482014112601T1_xxx". We will cut these seven images using the menu "Raster > Extraction > Clipper ...". In the window that opens, we select the first image to be cut "xxx_band1.tif", in "output file" we name the resulting raster "singur_2014-11-26_B1.tif" and we choose where to store it. For the "Clipping Mode", we choose "Mask layer" and we select the shapefile of boundary that will allow us to crop our image. We select the option "Crop the extent of the target dataset to the extent of the cutline" (Figure 3.1).

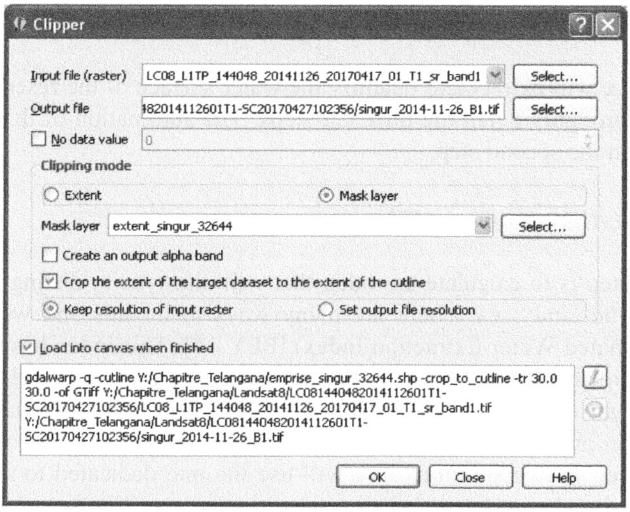

Figure 3.1. *Cropping of the Landsat images according to the extent of the study area*

After executing the module, cut-out strip 1 appears in the QGIS interface (Figure 3.2). We repeat this operation to cut out the remaining 6 bands that we call "singur_2014-11-26_B2.tif" up to "singur_2014-11-26_B7.tif".

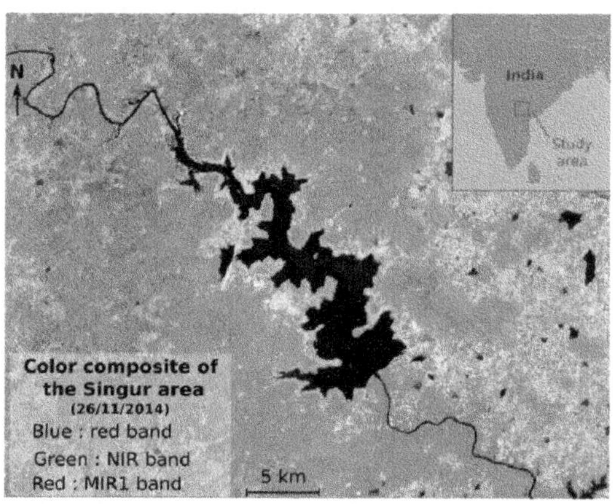

Figure 3.2. *Landsat images cropped according to the study gauge (17°24'N and 78°28'E, Singur Reservoir, Telangana State, India). The reservoir of the Singur appears in dark on this image. For a color version of the figure, see www.iste.co.uk/baghdadi/qgis4.zip*

For the rest of this chapter, we repeat this operation for the other 4 scenes.

3.3. Extraction and quantification of the Singur reservoir area

Initially, we will extract and quantify the water surface of the reservoir manually in order to thoroughly detail the different steps. The automation of this channel will be presented in the second step.

3.3.1. *Calculation of the AWEI Index*

The first step is to calculate an index that will allow us to distinguish the water bodies from the land area. There are many ways to do this and we will use the AWEI (Automated Water Extraction Index) [FEY 14]. This index has the advantage of transforming the pixels corresponding to water into positive pixels and the land pixels into negative pixels. This threshold of 0 is known to be very robust, allowing us to create an automatic extraction of water surfaces that we will see later. This index is derived in two solutions. We will use the one dedicated to areas of plains and flat areas that do not have clean or dropped shadows. The formula to be applied is as follows:

$$\text{AWEInsh} = 4 \times (\text{G-MIR1}) - (0.25 \times \text{NIR} + 2.75 \times \text{MIR2}) \qquad [3.1]$$

where 'G' is the green band (B3 of Landsat 8), 'NIR' is the near infrared band (B5) and MIR1 (B6) and MIR2 (B7) are the two middle infrared bands.

To calculate this index we use the "Raster Calculator..." found in the "Raster" menu. We enter the formula to calculate in "Raster calculator expression" and we save the result as "awei_2014-11-26.tif" (Figure 3.3).

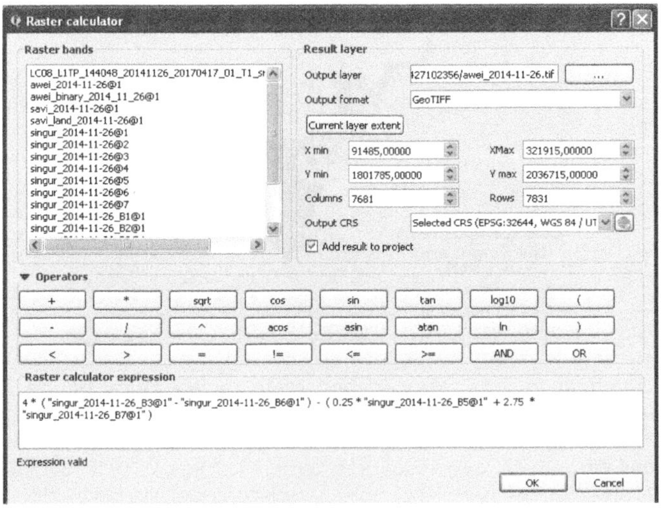

Figure 3.3. *Calculation of the AWEI index via the raster calculator*

Once the calculation is complete, the result will appear on the screen. After adjusting the style, that is to say the appearance of the layer, we obtain the raster presented in Figure 3.4. The pixels in turquoise are the pixels with positive values, the others have negative values. Water surfaces appear to be turquoise: we recognize our reservoir, isolated lakes and stretches of streams.

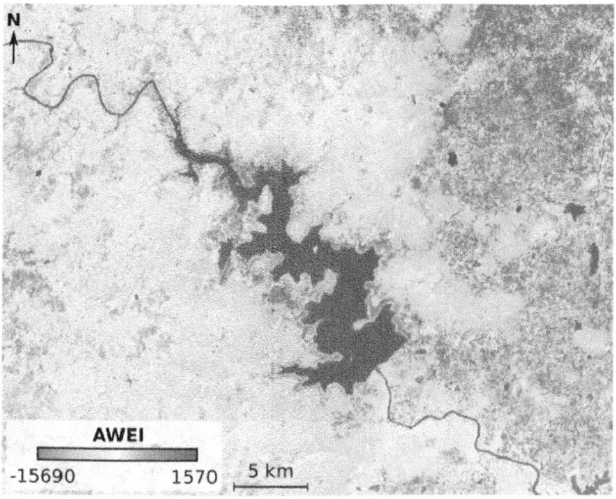

Figure 3.4. *AWEI for November 26th, 2014. The studied reservoir appears in turquoise in the center of the image. For a color version of the figure, see www.iste.co.uk/baghdadi/qgis4.zip*

We now have a raster from which it is easy to separate water surfaces from land surfaces. All positive pixels are water pixels, all negative pixels are earth pixels. Of course, the pixels located at the interface can oscillate around 0 and be misclassified. This uncertainty is one of the limitations of this approach.

3.3.2. *Construction of the water–land binary raster*

From the previous raster "awei_2014-11-26.tif", we will construct a binary raster where the water surfaces will be coded as 1 and the others as 0. We again use the raster calculator. This time, in the framework devoted to the expression to evaluate, we enter the following formula:

"awei_2014-11-26@1" > 0 [3.2]

And we save the result under the name "awei_binary_2014_11_26.tif". The following (Figure 3.5) appears in the main window.

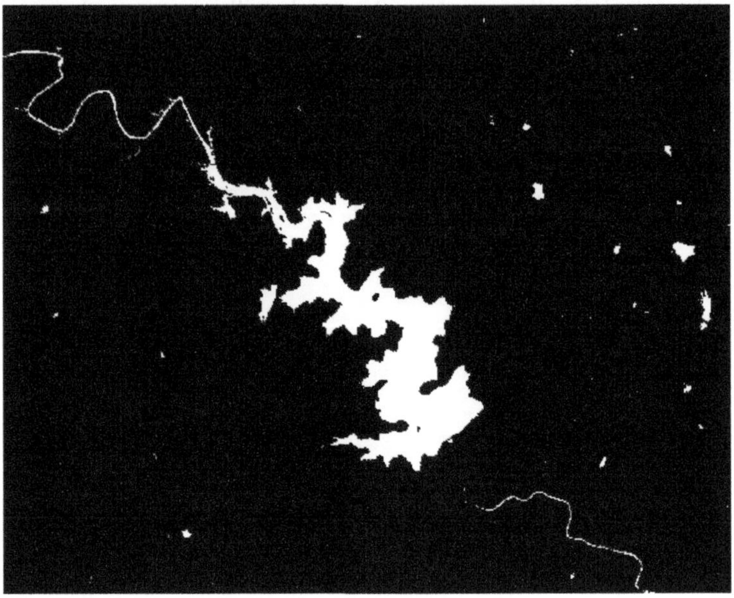

Figure 3.5. *AWEI binary raster. In white, the pixels have a value of 1 (water) and in black the pixels at 0 (earth)*

From a scientific point of view we can raise the question of the upstream limit of the reservoir. This work being on a river, it is difficult to say exactly where the mouth of the upstream river lies in the reservoir. For this chapter we will put aside these considerations to focus on the practical side.

3.3.3. *Vectorization of the binary raster*

Our goal is to know the water surface of the reservoir. If we want to work on attributes, it is easier to work with vector data. This is why we will vectorize our binary raster before continuing the analysis.

To do this we go to the menu "Raster > Conversion > Polygonize (Raster to Vector)...". In the window that appears, we enter the raster to be converted "awei_binary_2014_11_26.tif" and we specify where the shapefile corresponding to the conversion we call "water_land_2014-11-26.shp" is saved (Figure 3.6).

We obtain a "polygon" shapefile. The entities corresponding to the water surfaces have a "DN" attribute of 1 and the others of 0. We can check this fact by looking at the attribute table.

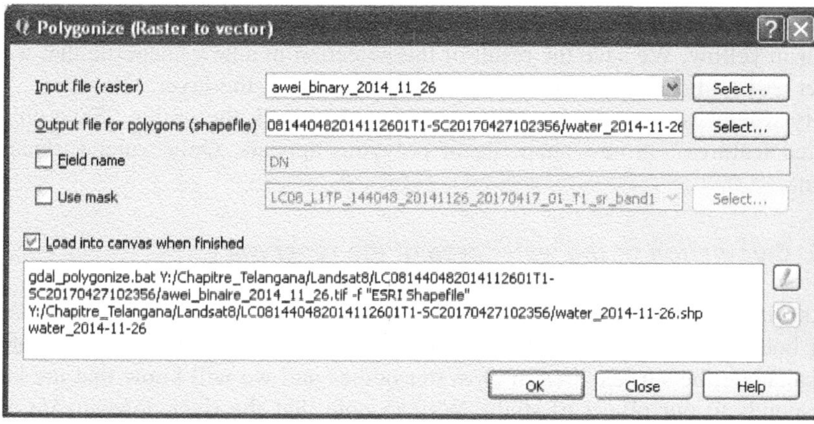

Figure 3.6. *Polygonization of the binary raster of the AWEI index*

3.3.4. *Selection of water polygons*

The next step is to select only polygons from water surfaces, that is, polygons with a "DN" attribute of 1. We use the "Select entities using an expression" tool (the icon looks like a rounded "E" on a yellow square in the toolbar). The vector layer on which we wish to make the selection must first be selected (highlighted in blue) in the layer panel. The query we use is simple ("DN" = 1) and we enter the left panel of the selection window (Figure 3.7).

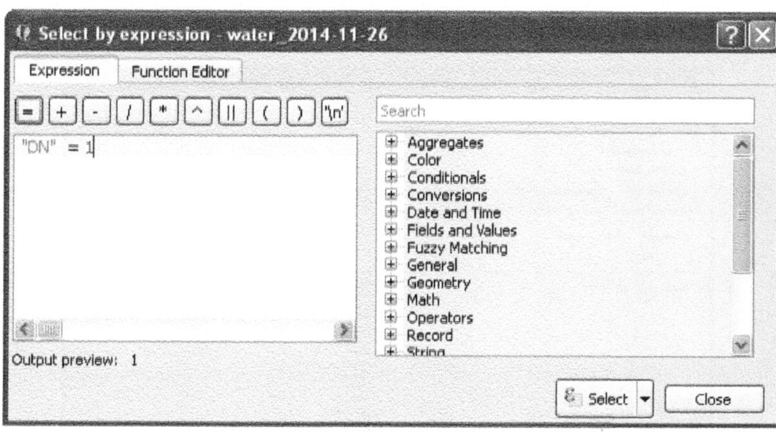

Figure 3.7. *Selection of polygons corresponding to water surfaces by using the formula* "DN"=1

At the end of the selection process, the entities corresponding to water surfaces appear in yellow. We save the result of this selection in a new shapefile that we call "water_2014-11-26.shp". To do this, we right click on the layer, "Save as ...", we specify where we want to save the result and we check the option "Save only the selected features". A new shapefile of polygons appears. Only water surfaces are present.

3.3.5. *Calculation of the water area of the reservoir*

As we see on the shapefile of water surfaces, the Singur reservoir is not the only water body in our study area. On the other hand, it is clearly the largest. Thus, we will calculate the surface area of all water bodies and we will know that the largest corresponds to our object of study. We can note that the river downstream of the reservoir is confounded with the reservoir, which will overestimate our reservoir area. We will see how to overcome this artifact in the next part dedicated to automating the process.

We calculate the surface area of water bodies via the "Field Calculator". Our layer "water_2014-11-26.shp" has to be selected and we click on the icon in the form of an abacus ("Open the field calculator") of the toolbar. In the window that appears (Figure 3.8), we create a new attribute that we call "area". This field is of the type "Decimal number (real)" with a length of 15 and an accuracy of 6 (length and accuracy are at the discretion of the user). In the "Expression" panel, we enter "$area", also accessible via the "Geometry" drop-down menu of the window's central panel. The result will be expressed in units of the projection used in the project, namely square meters.

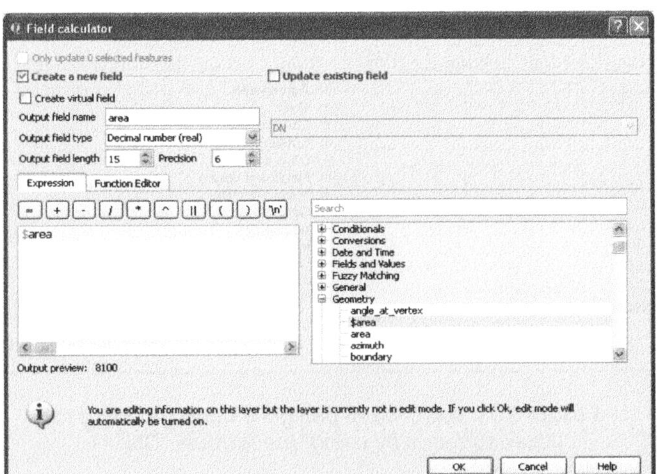

Figure 3.8. *Calculation of the surface area of water bodies via the Field Calculator*

At the moment of calculation, the layer automatically switches to "Edit" mode, which is why it appears in red. To exit this mode and save our area calculation, we must exit this mode by right clicking on the layer, selecting "Toggle editing mode" and saving changes.

When we open the attributes table of our layer: "water_2014-11-26.shp" we find a new column, "area", storing the square meters of each water surface. We click on the "area" column to sort it in descending order. The first line corresponds to the Singur reservoir, when we select it, the entity corresponding to the reservoir is also selected (Figure 3.9). Its surface area is 68,559,300 m^2, or 68.6 km^2.

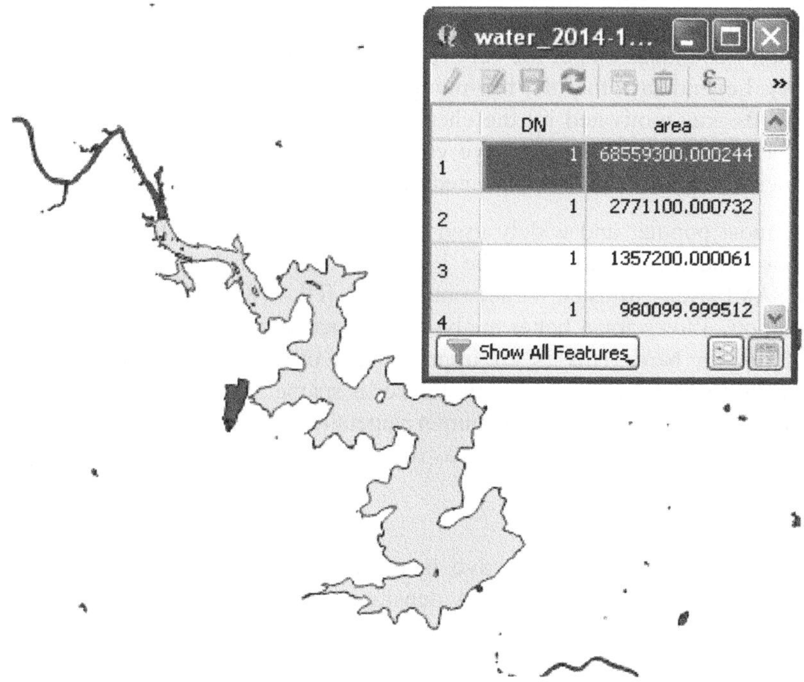

Figure 3.9. *Correspondence between the largest body of water and the entity corresponding to the Singur reservoir. For a color version of the figure, see www.iste.co.uk/baghdadi/qgis4.zip*

We have just developed a sequence of treatments to calculate the area of the Singur reservoir at a given date thanks to the analysis of a Landsat 8 image. The next point is to characterize the state of vegetation in the immediate vicinity of the reservoir for that same date. The next section deals with this point.

3.4. Characterization of vegetation

From a scientific point of view, it would first be necessary to define a kind of zone of influence of the reservoir. Indeed, is it appropriate to compare the area of the reservoir with the vegetation within 5 or 10 km? Or more? This area of influence can be very difficult to define and justify. In our case, this area is arbitrarily the one of our study area: we focus more on the practical side.

3.4.1. *Choosing an indicator of the state of vegetation*

As we work from satellite images we use an index dedicated to the characterization of vegetation. There are many of them. The monitoring of vegetation was indeed one of the very first applications of satellite imagery (after the weather). Legend says that the beginning of the American Landsat program in the early 1970s was motivated by the checking and evaluation of crops in the Soviet Union. This assessment would have enabled the US to more closely analyze the global cereal markets and the associated geopolitical issues.

The most popular and widely used index is the NDVI (Normalized Difference Vegetation Index), developed in the 1970s. In our example, we will use an index called SAVI (Soil-Adjusted Vegetation Index) [HUE 88]. This index, by its very construction, allows for a better analysis of vegetation when large fractions of the ground remain bare. This is an improvement of the NDVI that some authors recommend to use when the vegetation cover is sparse. However, our study region is semi-arid, so we can expect to encounter vegetation cover which is very variable in time and space. The SAVI formula is as follows:

$$SAVI = \frac{(1+L)(NIR - R)}{NIR + R + L} \qquad [3.3]$$

The coefficient L is a corrective factor which makes it possible to take into account the bare soil surrounding low density vegetation cover. L is generally set to 0.5.

In this part, we will calculate the SAVI for our zone and we will "mask" it with the water surfaces that we have previously extracted. This masking will allow us to only calculate relevant vegetation statistics on the land part of our domain.

3.4.2. *Calculation of the SAVI on the study area*

To calculate the SAVI, we use the raster calculator (see section 3.3.1). Based on the correspondences between spectral bands and Landsat 8 tape numbers, we enter the following formula in the "expression" panel of the calculator:

("singur_2014-11-26_B5@1" – "singur_2014-11-26_B4@1")/("singur_2014-11-26_B5@1" + "singur_2014-11-26_B4@1" + 0.5) * (1+0.5) [3.4]

We name the result "savi_2014-11-26.tif". After adjusting the style of the layer, we obtain the following raster (Figure 3.10).

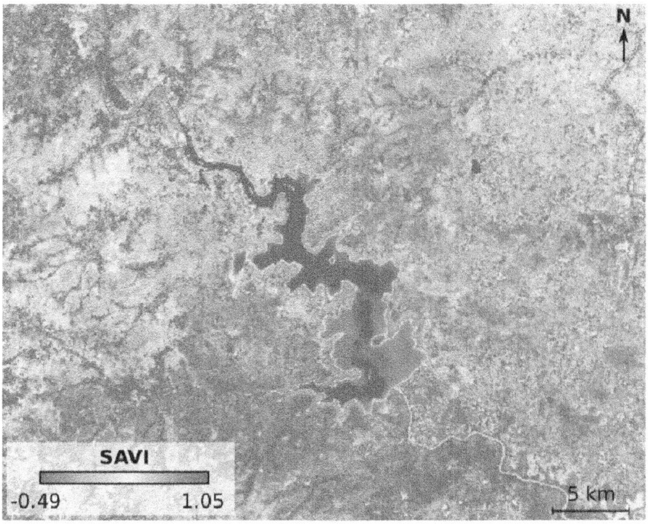

Figure 3.10. *SAVI near the reservoir of Singur. For a color version of the figure, see www.iste.co.uk/baghdadi/qgis4.zip*

Most of the negative values (blue in the previous figure) correspond to open water zones. We recognize the shape of the reservoir (blue). The higher the SAVI is, the greater the phytosynthetic activity of the vegetation. This can be interpreted as "healthy" vegetation. Here, this phenomenon is reflected by the intensity of red. The darker red a pixel is, the more the pixel is covered with healthy vegetation. Conversely, the closer to white a pixel is, the less this pixel is covered with vegetation.

3.4.3. *Creating a land–water mask*

In order to maintain consistency in our analysis, we will free ourselves from open water surfaces in the calculation of SAVI statistics. It is indeed irrelevant to look at the state of vegetation on water bodies. Even if aquatic vegetation can develop, it would not be interesting in our study. We will thus mask the water bodies using the binary raster of the AWEI index "awei_binary_2014-11-26.tif" created in the previous section. The aim is to obtain the SAVI values for the land parts and "no data" values for the parts in water.

On the AWEI binary raster we have water surfaces coded as 1 and earth as 0. We start by subtracting 1 from this raster. We thus obtain the water surfaces coded as 0 and the ground as -1. We take the inverse of this raster (1/raster) and we get “no data” for water surfaces (due to the division by 0 which is impossible) and -1 for the earth. At this stage it is enough to multiply this raster by -1 to obtain the water as “no data” and the earth as 1. Finally, we apply this mask to the SAVI raster simply by multiplication. On the land pixels we obtain SAVI and on the water bodies “no data” * SAVI which gives “no data”. We apply this simply operation with the raster calculator. The following formula must be entered in the expression panel:

(- (1 / ("awei_binaire_2014_11_26@1" - 1))) * "savi_2014-11-26@1" [3.5]

We obtain the following raster (Figure 3.11).

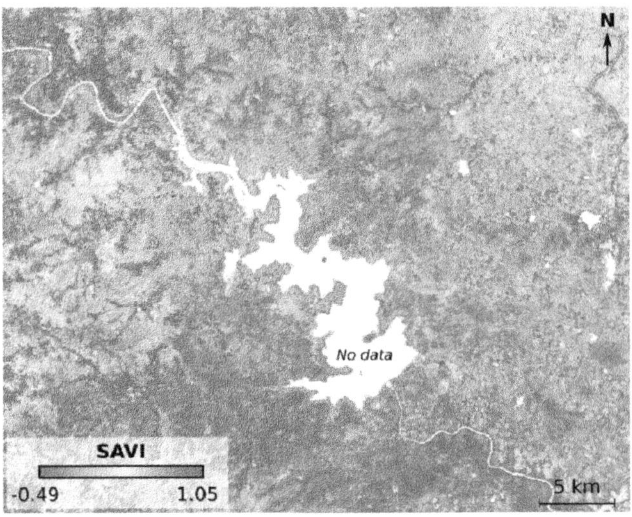

Figure 3.11. *SAVI masked to only keep the land pixels. The water surfaces appear here in white but they are actually “no data”. For a color version of the figure, see www.iste.co.uk/baghdadi/qgis4.zip*

3.4.4. *Statistics of the SAVI land surface index*

We now have a SAVI raster on land surfaces, perfectly in line with our water surfaces for the given date. This SAVI is a good indicator of the vegetation status in our study area. We now want to summarize this indicator by calculating the average SAVI of the area. This calculation is done by simply using the “Raster layer statistics” tool in the “Processing Toolbox > QGIS geoprocessing > Raster tools” panel. In the window that opens, we specify which raster we want to summarize “savi_land_2014-11-26.tif” and we name the file that will contain the statistics “stats_savi_land_2014-11-26”. In the end, the mean SAVI value is 0.69.

We have detailed the steps allowing us to calculate the area of the Singur reservoir for a given date and to characterize the surrounding vegetation for the same date. In order to process several dates in order to build time series, we will see in the next section how to automate these steps.

3.5. Automation of the processing chain via the construction of a QGIS model

In the two previous sections we have seen the different steps needed to answer our initial question in detail. The procedure is quite long and it can be painful to apply for different dates. The automation of this procedure will enhance the efficiency of the method. The procedure will become semi-automatic, it will still only be possible to process Landsat scenes one after the other and a post-processing step will always be necessary. A complete automation is quite possible but it would involve the writing of scripts (in Python for example) that would take us away from the aim of this chapter.

This automation will be done using the QGIS model builder. In this constructor, it is possible to graphically create the sequence of algorithms that we wish to realize.

3.5.1. *Model setting*

For this section, we start from a blank QGIS project in which we load our 7 Landsat images corresponding to the 7 spectral bands of the scene to be analyzed. We can use the scene we have already dealt with, that of November 26, 2014.

We can build our own templates from the "Processing Toolbox > Models > Tools > Create new model". A "Processing modeler" window opens (Figure 3.12).

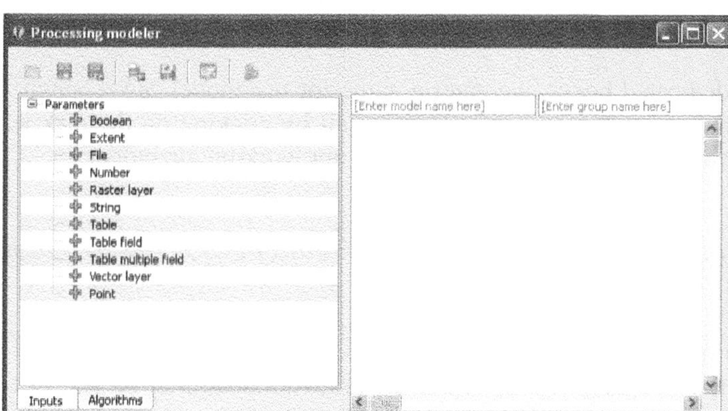

Figure 3.12. *The processing modeler window*

In the left panel, we have a list of all types of parameters that we can use in our model. For this example, we will use “raster layer” and “vector layer”. If we click on the “Algorithms” tab at the bottom of this panel, we get the list of algorithms we can call upon in our model (Figure 3.13).

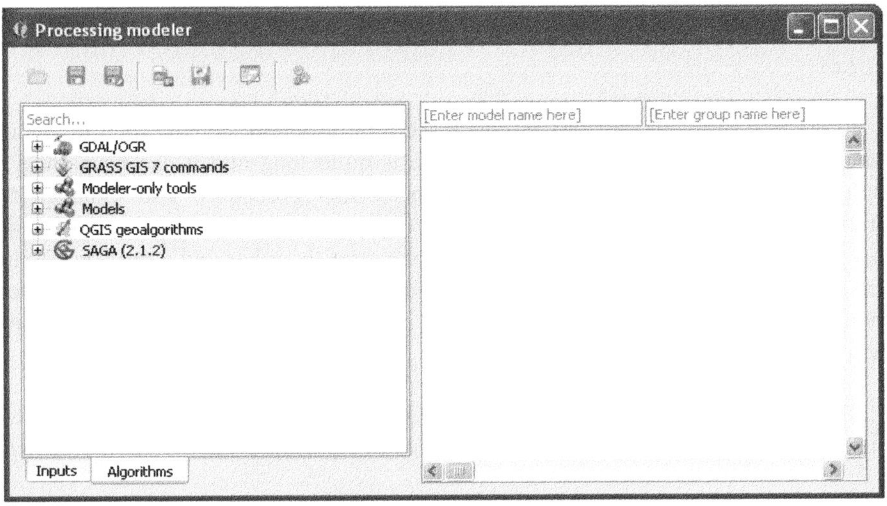

Figure 3.13. *List of algorithms that we can call upon in our model*

We can call the modules GRASS GIS, SAGA GIS, GDAL, QGIS ...

In the right panel, we will drag and drop the parameters and modules we want to chain. This method is completely graphic, there is no need to code anything. Although it appears easy to use this graphic method, there are some limitations inherent to this approach (no iterations, no logical tests, no fine parameterization ...)

In the dedicated boxes, we enter a name for our model “awei_singur_en” and a name for the group “singur_en”. We save all by clicking on the disk icon and we name our model “singur” which we store in our working directory.

3.5.2. *Construction of the chain of treatments for the extraction of the reservoir*

In this model, we make a small modification compared to the theoretical algorithm that we have just formulated. As mentioned, after an AWEI calculation, the positive pixels correspond to water surfaces and the negative pixels correspond to land surfaces. However, in our case, the reservoir studied may be very shallow and very turbid at certain dates, especially during dry periods. These characteristics have a non-negligible impact on the corresponding spectral response of the surface to water. Thus, we do not set the threshold at 0 but at -1000. Any pixel greater than -1000 will be considered as a water pixel. Let us not forget that these images are coded from 0 to 10000 rather than from 0 to 1. A threshold at -1000 therefore corresponds to "true" reflectance at only -0.1. It is often necessary to find compromises when we try to carry out tasks automatically.

The chain of treatments allows us to extract the reservoirs and to characterize the vegetation for a given date. To process the four dates, it is necessary to rotate this chain of treatments four times.

3.5.2.1. *Definition of the first link in the model*

The first step in this chain is to calculate the AWEI index on our region. As we saw in the previous section, the calculation of this index takes 4 different rasters as input, corresponding to the green and near infrared bands and the two bands of middle infrared. We start by dragging/dropping four "Raster layers" from the "Inputs" panel to the main panel. Each time we are asked to name the raster. We call them successively "green, nir, mir1 and mir2" (Figure 3.14).

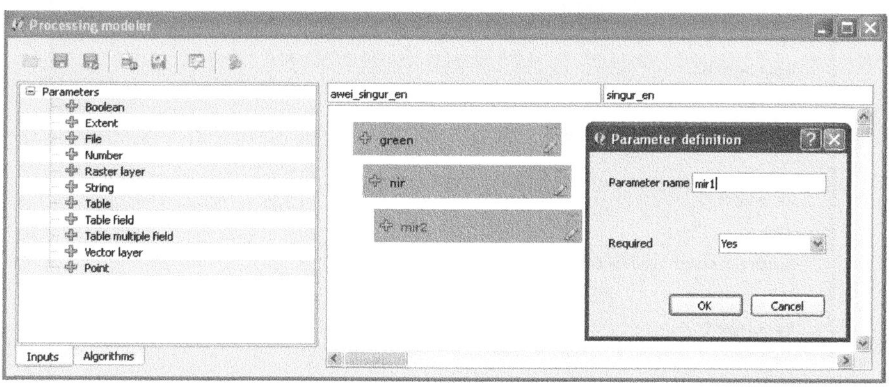

Figure 3.14. *Import of the bands used to calculate the AWEI index*

Once these bands are loaded, we derive the AWEI index. We add an algorithm to our model. We must use a model equivalent to the raster calculator. In our case, we choose the module "GDAL / OGR> [GDAL] miscellaneous > Raster calculator" (Figure 3.15). A setting window opens. We indicate the rasters we want to work on and the operations we want to carry out. As a description, we can write "AWEI Calculation". Then we configure the first raster (the first layer) that we want to use.

For the algorithm, this will be "input layer A". In our case, this is the "green" raster. We do the same for layers B, C and D which are respectively the rasters "nir", "mir1" and "mir2". Then we write the expression to be evaluated in the line "Calculation in gdalnumeric syntax ...". Layers must be called by their letter and not by name. The calculation of the AWEI is thus done with the following formula, without spaces:

$$4*(A-C)-(0.25*B+2.75*D) \quad [3.6]$$

At the "Calculated" line (at the bottom of the "Raster calculator" window) we specify a generic name for the product created, "awei". The other options may retain their default values.

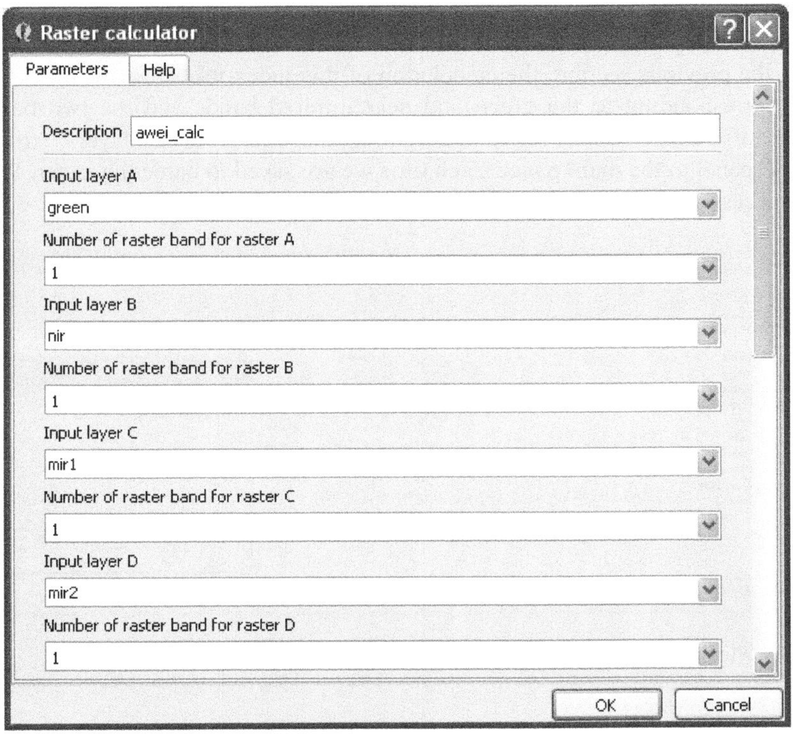

Figure 3.15. *Parameterization of the "Raster calculator" algorithm to calculate the AWEI index within the model*

At the end of this step, our model resembles the following (Figure 3.16). It is important to regularly save his model.

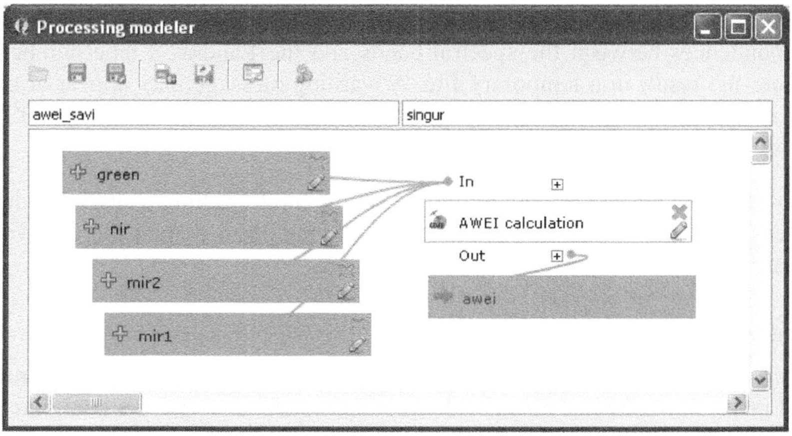

Figure 3.16. *The first step of our model: calculation of the AWEI index*

We can already test our model by running it. To do this, simply click on the gear-wheel icon in the toolbar of the "Processing modeler" window. At launch, a window for setting the inputs to be used opens. This window bears the name of our model (Figure 3.17).

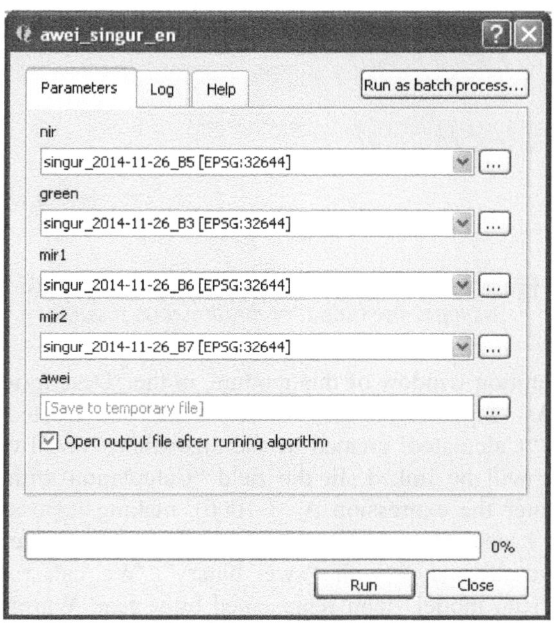

Figure 3.17. *Launch test of the first step of our model*

Be careful, the rasters are not in order. We take care to observe the correct correspondences between the spectral bands and the Landsat 8 tape numbers. We can store the result in a temporary file. A warning message may appear at runtime but without affecting the result. The AWEI raster appears in the QGIS main window.

Now that we have seen how to build a simple model calling a single module, we will see how to link the various modules allowing us to extract the reservoir and calculate the SAVI to characterize the vegetation.

3.5.2.2. *Linking modules within the model*

As we saw in the previous section, the second step consists of transforming our AWEI raster into a binary raster with the water pixels at 1 and the land pixels at 0. The insertion of this module is done simply by taking the output of the module used to calculate the AWEI as input (Figure 3.18). The module we use is still the module "GDAL / OGR > [GDAL] miscellaneous > Raster calculator".

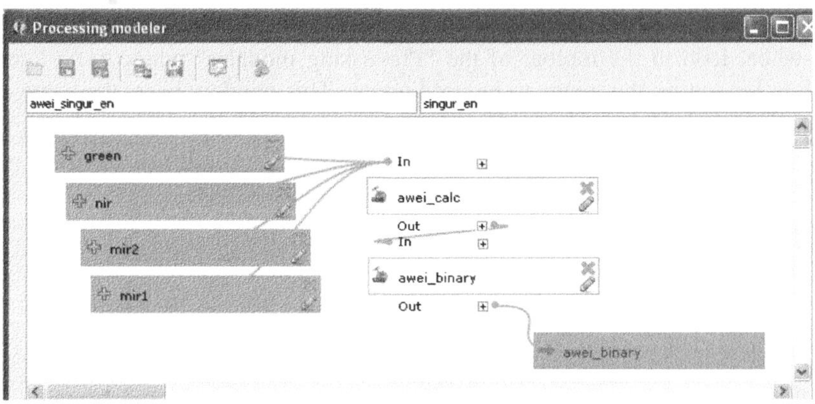

Figure 3.18. *Sequence of the second module taking as input the output of the previous module*

In the configuration window of this module, in the "Description" field we enter "awei_binary". As "Input Layer A", we select the one created by the previous module, namely "'Calculated' created by the algorithm 'AWEI calc'", in this way the two modules will be linked. In the field "Calculation with the gdalnumeric syntax ...', we enter the expression A> (-1000), making it possible to obtain the water pixels in 1 and the others in 0. In order to visualize the result, field "Calculated <OutputRaster>" with "awei_binary". We can run our model by clicking on the "Run model" icon represented by a gear. Warning messages may appear but will not affect the end result. The binary raster appears as an end product. It may be necessary to change the style to visualize it.

3.5.2.3. *Writing the complete model*

The various modules to be connected are described in the following lines. Note that for the calculation of the SAVI, it is necessary to add the parameter "Raster layer" corresponding to the red band (band 4). We have just completed the first two steps. We have only to complete our model from the step where we vectorize the binary AWEI raster. If we do not want to output the intermediate products, we leave the "Calculated <OutputRaster>" field blank.

AWEI calculation:

– GDAL Raster calculator

- Purpose: awei_calc
- Description: awei_calc
- Inputs: green (B3), nir (B5), mir1 (B6), mir2 (B7)
- Output: AWEI with water> 0 and ground <0

– GDAL Raster calculator

- Purpose: separation of water (1) and soil (0)
- Description: awei_binary
- Input: AWEI produced by the previous module
- Formula: A> - (1000)
- Output: AWEI binary with water in 1 and earth in no data

– GDAL Polygonize (raster to vector)

- Purpose: to vectorize the binary raster
- Description: awei_binary_vect
- Input: AWEI binary produced by the previous module
- Output: shapefile with water polygons in 1 and ground in 0

– Geoprocessing QGIS Extract by attributes

- Purpose: to extract water polygons (polygons with 1)
- Description: extract_water
- Input: the binary shapefile produced by the previous module
- Selection attribute: DN

- Operator: =

- Value: 1

- Output: water surface polygons

– QGIS Geoprocessing Field Calculator

- Purpose: To calculate the area of each water polygon

- Description: calc_area

- Input: the polygons of the water surfaces produced by the previous module

- Parameter:

1) Name of the field: area

2) Field Type: Float

3) Length of field: 10

4) Field Accuracy: 3

5) Create a new field: yes

6) Formula: $area/1000000 (to have km^2)

- Output: the same polygons but with an attribute corresponding to the area of each of them expressed in square kilometers. We retain the result shapefile and we call it "water_area".

– QGIS Geoprocessing Basic statistics for numeric fields

- Purpose: to find the maximum water area that corresponds to the Singur (because we know that the Singur is the largest area of water in the image)

- Description: area_singur

- Input: the layer of water polygons with the surface area produced by the previous module

- Output: summary html file, the maximum value corresponds to the Singur. We keep this file under the name "stats_singur".

SAVI calculation:

– GDAL Raster calculator

- Purpose: Calculation of SAVI without regard to water areas, (or SAVI calculation for land)

- Description: savi_calc

- Input: NIR (B5, seen as layer A by the module), red (B4, seen as layer B by module), and AWEI (seen as layer C by module)

- Formula: (C<-1000)*((A-B)/(A+B+0.5)*(1+0.5)). We multiply the SAVI by (C<-1000) in order to calculate it only for land pixels.

- Output: the SAVI raster

– QGIS Geoprocessing Basic raster layer statistics

- Purpose: To calculate the average SAVI for the study area

- Description: stats_savi

- Input: the SAVI raster created by the previous module

- Out: a summary html file. We store this file as "stats_savi".

In the end, we obtain the following QGIS model (Figure 3.19). We have 5 input rasters (Green, NIR, MIR1, MIR2 and Red, respectively B3, B5, B6, B7 and B4), an outline shapefile ("water_area") containing the polygons of the water surfaces with their associated surfaces, and two html files containing the statistics of water areas ("stats_singur") where the largest corresponds to the Singur reservoir and the second file ("stats_savi") contains the statistics of the SAVI of the study area.

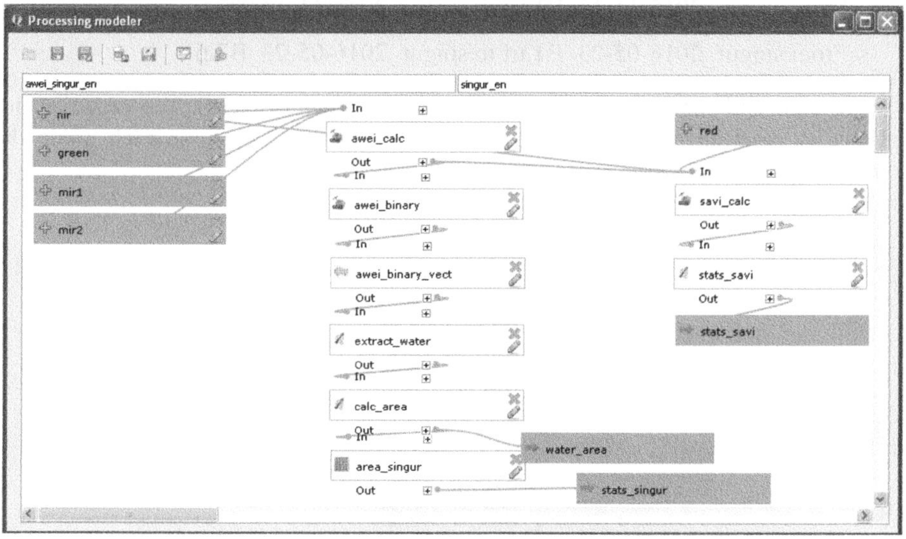

Figure 3.19. *The complete model with 5 input rasters, a shapefile and two html files outputs and 8 modules that are connected. For a color version of the figure, see www.iste.co.uk/baghdadi/qgis4.zip*

We can now test our model on our first Landsat 8 scene on November 26, 2014. We remind you that our model appears in the "Toolbox" panel in the "Models" menu. We must be careful to respect the concordance between spectral bands and raster names when launching the model. We can name the outputs "water_area" as "water20141126.shp", "stats_singur" as statsWater20141126.html" and "stats_savi" as "statsSavi20141126.html". Once the module is executed, the shapefile of the water surfaces is displayed in the main window and a new window opens the two html files within QGIS. We find the surface of the Singur (maximum value) in the summary file of the water surfaces. The value is a bit different from the previous one because we have changed the threshold of the AWEI in the model. The average value of the SAVI of the region is in the other html file.

3.5.2.4. *Running the model on other scenes*

The interest of this type of model is to be able to execute these processes in a single operation on different scenes corresponding to different dates. We will then run our model on our three other images corresponding to the dates of May 21, 2015, November 13, 2015 and May 23, 2016. These three other scenes must first have been cropped according to the boundaries of the study area. The reader can do this themselves in the same way as explained in the section "Preparing the data" or directly using the already cropped scenes provided. Each scene is named as follows:

– From singur_2015-05-21_B1.tif to singur_2015-05-21_B7.tif

– From singur_2015-11-13_B1.tif to singur_2015-11-13_B7.tif

– From singur_2016-05-23_B1.tif to singur_2016-05-23_B7.tif

We can load the 7 spectral bands of each scene into QGIS. We organize our project by creating "groups" in the "Layers" panel. We can name each group the name of the corresponding scene (Figure 3.20).

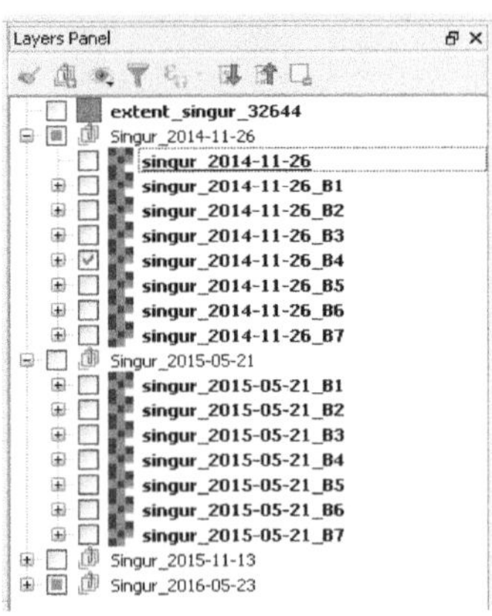

Figure 3.20. *Organization of the layers for the execution of the model*

We will now run our model three times, once on each scene corresponding to our three dates. We start by executing our chain of treatments on the stage of May 21, 2015. For this we execute our model with the following parameters (Figure 3.21).

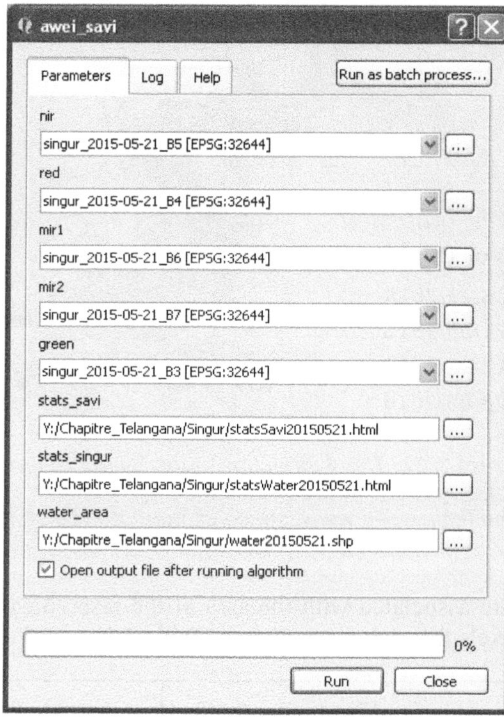

Figure 3.21. *Execution of the model for the second date of our study*

We name the outputs as follows:

– water20150521.shp

– statsSavi20150521.html

– statswater20150521.html

We proceed in the same way for the following two dates, following the same pattern to name our outputs. We obtain the following (Figure 3.22).

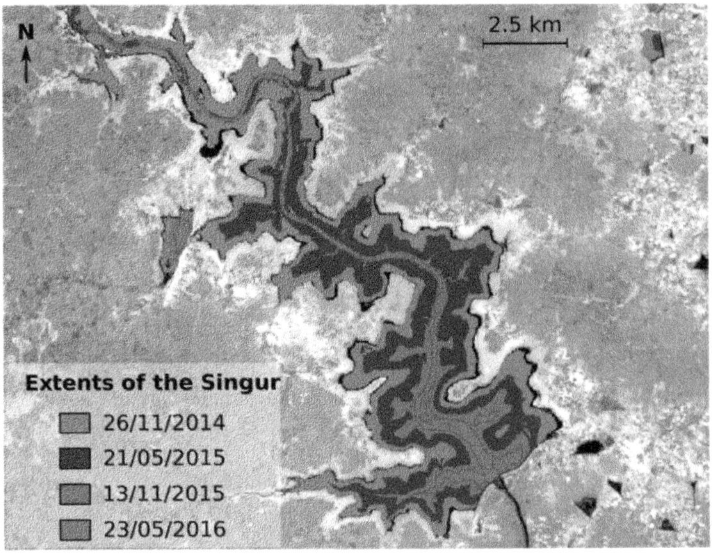

Figure 3.22. *The 4 stretches of the Singur reservoir corresponding to the 4 processed dates. For a color version of the figure, see www.iste.co.uk/baghdadi/qgis4.zip*

These extents are associated with the area of the reservoir and the mean SAVI value of the land pixels for each piece of data (Table 3.1).

Date	Singur area (km²)	Average SAVI on the zone
November 26, 2014	75.70	0.65
May 21, 2015	39.56	0.47
November 13, 2015	13.99	0.67
May 23, 2016	3.00	0.38

Table 3.1. *Area of the Singur reservoir at each date and the average values of the corresponding SAVI*

It is obviously not relevant to draw conclusions only on these dates, but it appears that there is no connection between the area of the Singur reservoir and the surrounding vegetation. Vegetation shows two peaks of activity in November for both years. While the maximum area of the reservoir is not necessarily in

November. It would be necessary to study a more representative sample of images in order to draw relevant conclusions.

3.6. Conclusions

In this chapter we have seen how to extract water surfaces from Landsat 8 satellite imagery and how to compare these surfaces to the state of the surrounding vegetation. We have developed an algorithm based on the AWEI index for the extraction of water surfaces and on the SAVI index for the characterization of vegetation. We were able to create our own model that automates our algorithm using the QGIS built-in model generator. The modules used in the model are not always exactly what we would use at first glance. It is sometimes necessary to use a roundabout method. It is also necessary to change certain parameters, such as the threshold on the AWEI, in order to best match the spectral responses that can fluctuate greatly from one date to the next.

3.7. Bibliography

[ADU 11] ADUSUMILLI R., LAXMI S.B., "Potential of the system of rice intensification for systemic improvement in rice production and water use: the case of Andhra Pradesh, India", *Paddy Water Environment*, vol. 9, no. 1, pp. 89–97, 2011.

[CEL 10] CELIO M., SCOTT C.A., GIORDANO M., "Urban–agricultural water appropriation: the Hyderabad, India case", *Geographical Journal*, vol. 176, pp. 39–57, 2010.

[FEY 14] FEYISA G.L., MEILBY H., FENSHOLT R. *et al.*, "Automated water extraction index: a new technique for surface water mapping using Landsat imagery", *Remote Sensing of Environment*, vol. 140, pp. 23–25, 2014.

[HUE 88] HUETE A.R., "A Soil-Adjusted Vegetation Index (SAVI)", *Remote Sensing of Environment*, vol. 25, pp. 295–309, 1988.

[PAS 18] PASSY P., THÉRY S., "The use of SAGA GIS Modules in QGIS", in BAGHDADI N., MALLET C., ZRIBI M. (eds), *QGIS and Generic Tools*, ISTE Ltd, London and John Wiley & Sons, New York, 2018.

4

Network Analysis and Routing with QGIS

4.1. Introduction

The purpose of this chapter is to present the various treatments offered by QGIS for network analysis and routing. After a quick definition of networks and graphs, we will discuss the rules for construction and verification of networks using GIS. Section 4.3 is devoted to the presentation of examples of hydrographic network development and analysis. Finally, section 4.4 consists of several practical works focused on a hydrographic network analysis and routing as part of a global project.

4.2. General notions

4.2.1. *Definition of a network*

A network is defined as a set of lines or elements that communicate or intersect. A more precise definition evokes a frame or structure composed of elements or points, often referred to as nodes (or vertices), connected by links (or lines, arcs or edges) ensuring their interconnection or interaction (source: Wikipedia). Graphs are abstract models permitting us to model networks. We distinguish the graphs by taking into account the direction of the links [BEA 10]. If the links are oriented, the relation goes in only one direction and is thus asymmetric, and the graph itself is said to be "oriented". We also separate simple graphs from other graphs. Graph theory defines a simple graph as a set of vertices without a loop (a link from a vertex to itself) and without multiple links (two vertices connected by several links). Hydrographic networks are represented by simple oriented graphs.

Chapter written by Hervé PELLA and Kenji OSE.

The Institut National de l'Information Géographique et Forestière (IGN) ROUTE 500® database is a good example of a network-based digital product. This database contains the entire classified road network (highways, national roads, departmental roads), characterized by a number of properties (road number, viability, importance of links, etc.) and supplemented by elements of the secondary road network and the main transport infrastructures (source: IGN® Route500). The organization and coherence of spatial objects, the description of the communication logic and the attachment of the 36,600 French communes that make up this network allow network traversal. This navigation can take several forms: finding the best route between two locations as well as various applications related to road transport (finding constrained city-to-city routes and their associated costs, transport optimization, multimodal transport and geomarketing applications).

4.2.2. *Network topology*

The topology of GIS is expressed by a set of rules dealing with relations between spatial entities. It is with respect to these rules that the topological coherence of the network is determined. This is a prerequisite for any form of spatial analysis.

Network consistency consists of checking:

– Overlay: the lines can intersect or cut each other, but they cannot have segments in common (see Figure 4.1(a)). In the case of common segments, the network consistency procedure removes one of the segments by splitting this segment into two distinct segments.

– Intersection: intersection must occur only at the ends, i.e. at the nodes (see Figure 4.1(b)). When two lines are intersecting, the procedure divides the two lines into two distinct segments.

– Dangles: the endpoint of a line that is not connected to another line is called a dangle (see Figure 4.1(c)). The dangles can be moved to place them exactly at the location of the reference geometry ("snap" operation). An exception is made in the case of hydrographic networks for the upstream nodes of the arcs located at the head of the watershed (headwater).

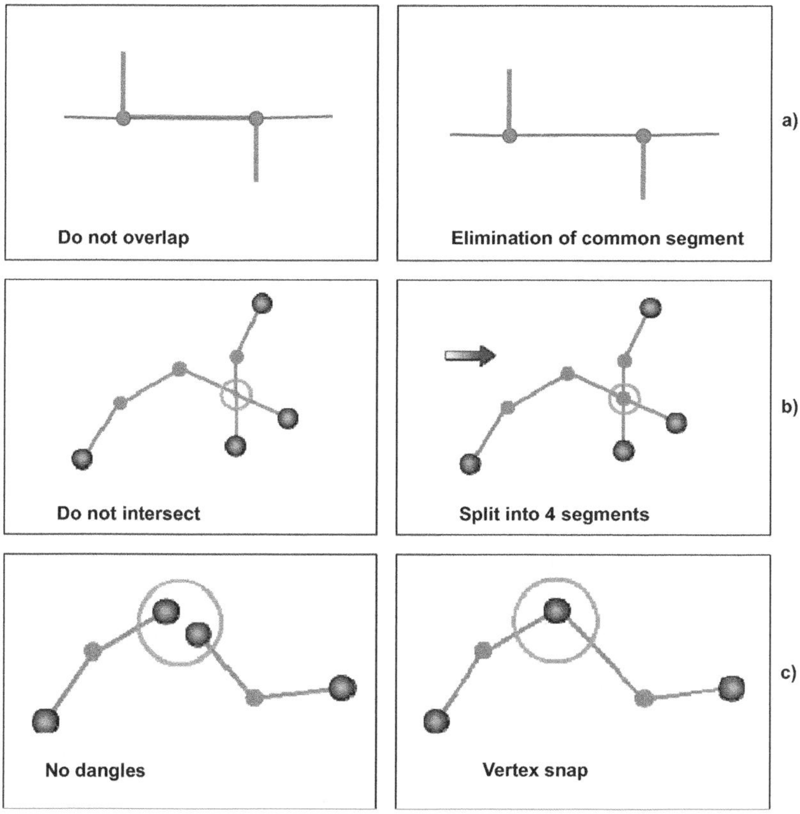

Figure 4.1. *Network[1] topological rules concerning a) superposition, b) intersection and c) dangle presence. For a color version of the figure, see www.iste.co.uk/baghdadi/qgis4.zip*

4.2.3. *Topological relationships*

In the case of networks, the notion of adjacency or contiguity implies that the spatial units have a common node. This notion permits us to know the arcs directly connected to the target arc. This non-metric topology only takes into account the order in which spatial units are moving away from the target unit. Adjacency is said to be of first order if the two entities are in contact (i.e., they have a node in common), second order if an entity is intercalated (see Figure 4.2), etc. For example,

1 http://www.portailsig.org/content/grass-gis-geometries-topologies-et-consequences-pratiques-vecteurs-rasters-volumes.

adjacency order occurs in transport to determine the number of changes required to move people from one location to another.

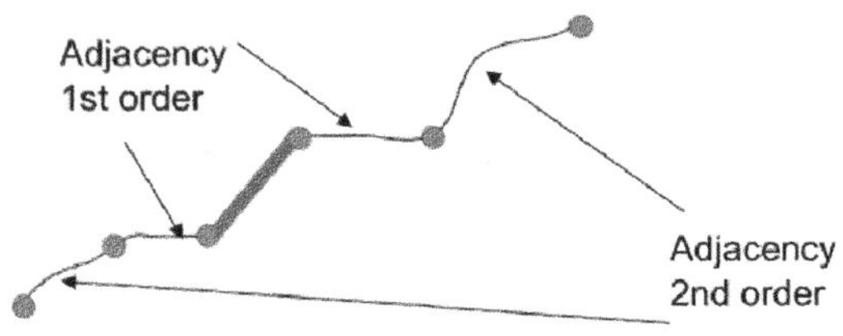

Figure 4.2. *Adjacency for a simple oriented network. For a color version of the figure, see www.iste.co.uk/baghdadi/qgis4.zip*

Connectivity expresses adjacency for linear networks. It can be oriented, as is the case for hydrographic networks. Connectivity then expresses the direction of flow of water through the network. The direction of flow can be carried by the direction of description of the arcs. In this case, the points forming the arcs are digitized from upstream to downstream (see Figure 4.3). From a source node, graph traversal is done by considering the neighboring vertices and thus using the properties of continuities between the different arcs.

The direction of flow can also be carried by the topology (see Figure 4.4). Each arc is then defined by its upstream node (First node or Fnode in the ArcInfo topology) and its downstream node (To node or Tnode in the ArcInfo topology). Two successive arcs share the same vertex (the "end" vertex of the upstream arc is the same as the "start" vertex of the downstream arc). In the same way, when two arcs flow together, they share the same "end" vertex. In this case, graph traversal can be done using only the topology without considering the geographical entities.

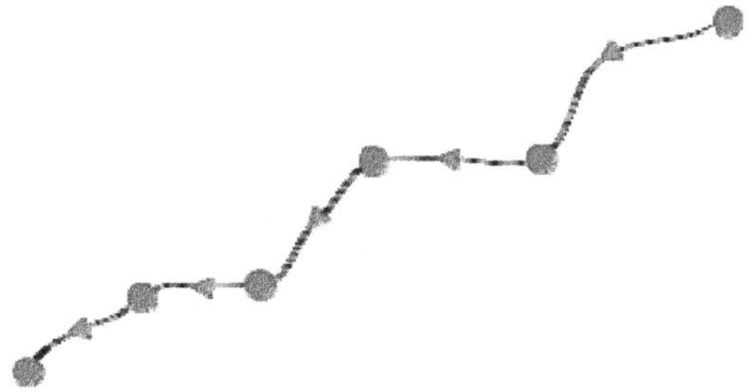

Figure 4.3. *The direction of flow of the arcs is carried by the direction of description of the nodes which compose them. For a color version of the figure, see www.iste.co.uk/baghdadi/qgis4.zip*

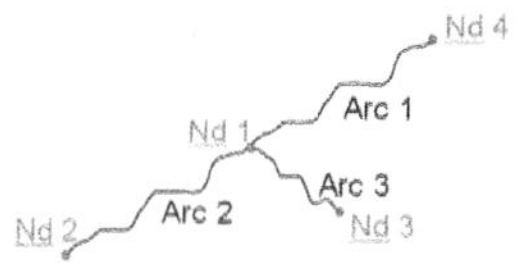

Arc	Start	End
Arc 1	4	1
Arc 2	1	2
Arc 3	3	1

Figure 4.4. *The direction of flow of the arcs is carried by the topology. For a color version of the figure, see www.iste.co.uk/baghdadi/qgis4.zip*

4.2.4. *Graph traversal: example of the shortest path (Dijkstra)*

The best-known graph traversal algorithm is probably the one developed by the Dutch computer scientist Edsger Dijkstra [DIJ 59]. This algorithm allows us to calculate from a directed and weighted graph the shortest path from a starting vertex to all the other vertices. The goal is to compute a subgraph listing the various vertices in ascending order of their minimum distance to the starting vertex. The algorithm begins by calculating the distance from the selected vertex to all the other vertices. From the nearest vertex (the vertex of least weight), it determines, once again, the distance for each of the arcs. And so on, until all the vertices have been covered. The shortest path is then that of lower cost (see Figure 4.5). In the case of a hydrographic network, the treatment is simpler, since it is necessary to traverse the network downstream from the starting node to the outlet.

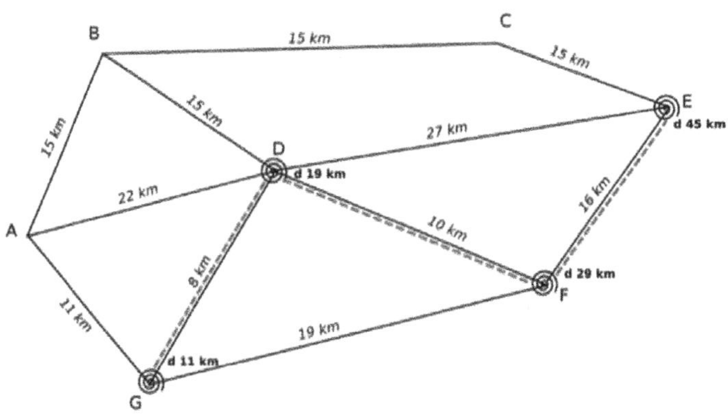

Figure 4.5. *Analysis of the shortest path (dotted line) by the Dijkstra algorithm from a weighted graph*

4.3. Examples of development and analysis of hydrographic networks

The development of GIS has made it possible to link environmental variables with the course of rivers. The need to take into account the influence of the upstream has gradually led to the development of tools and methods applied to this topic. The land use analysis of the river section watershed, the position of a river in the network, the influence of distance to source or sea for modeling existing fish populations or the modeling of the flow along the network are some examples [OBE 01].

The use of a hydrographic network allows us to integrate spatially the physical parameters related to topography (altitude, slope), geography (length of sections, watershed surface, etc.), climate (precipitation, temperature) and hydrology (liquid and solid flow) and to accumulate them downstream and thus provide useful data in the fields of hydrology, geomorphology or ecology.

For example, in order to allow resource inventory and stream management, a New Zealand River Environment Classification System (REC) was developed. This network provides a framework for the evaluation of impact studies, the definition of management plans, the development of monitoring programs and the interpretation of field data [SNE 02]. For example, it was used to analyze the contribution of snowmelt to the annual mean flow for rivers of order 4 and above (see Figure 4.6).

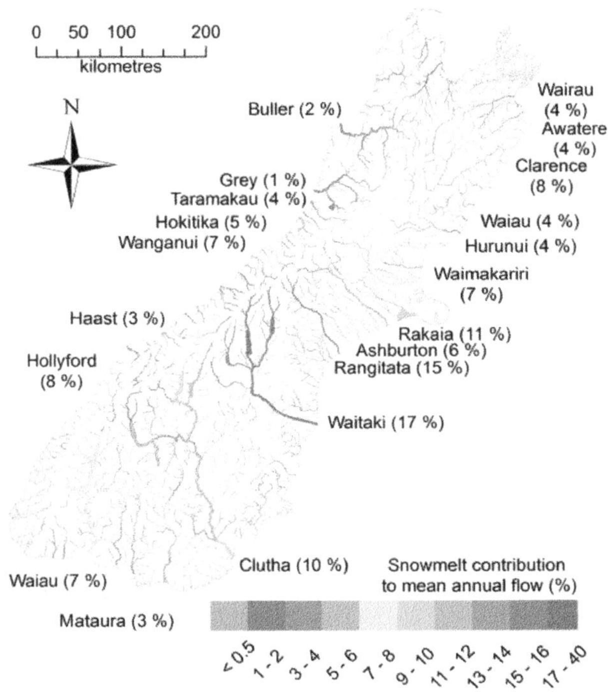

Figure 4.6. *Snowmelt contribution to mean annual flows for rivers of Strahler order 4 and higher. Percentages given in brackets are the estimated snowmelt contribution at the river mouths for a selection of larger rivers in New Zealand [KER 13]. For a color version of the figure, see www.iste.co.uk/baghdadi/qgis4.zip*

In France, a network of this type (theoretical hydrographic network (RHT) was also built with the initial aim of having a coherent topological network at the level of France and associating it with a set of environmental attributes (see Figure 4.7), which can be exploited by different scientific and operational actors [PEL 12]. The developed network is used in various projects such as nitrogen flux modeling [DUP 13], definition of a theoretical fragmentation indicator for aquatic environments (Onema1), estimation of water abstraction impacts on aquatic habitats [MIG 16], study of eel distributions and the role of obstacles to flow [JOU 12].

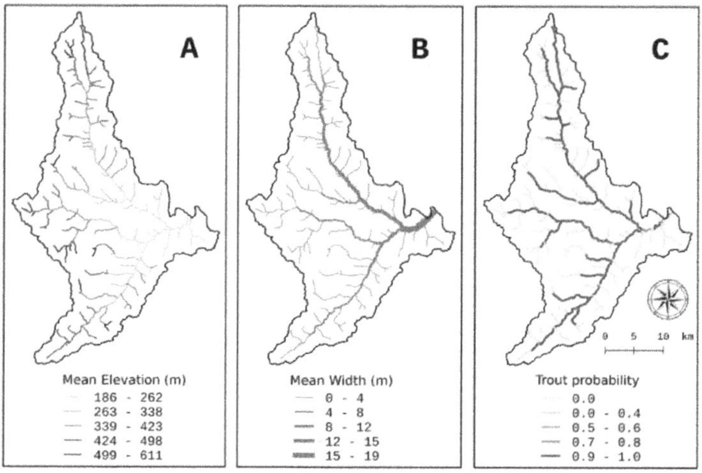

Figure 4.7. *Cartography of the (A) mean elevation, (B) average width and (C) the presence probability of trout on the RHT arcs for the Azergues watershed (France). For a color version of the figure, see www.iste.co.uk/baghdadi/qgis4.zip*

4.4. Thematic analysis

4.4.1. *Introduction*

The thematic analysis is a fictitious case developed by the authors with the sole aim of presenting a coherent approach within the framework of practical work. Apart from the RHT and the subdrainage basins, all the other spatial elements are fictitious.

The proposed problem is as follows:

> *A fishing federation observes the presence of the Louisiana crayfish in a drainage basin. This exotic species, which is expanding at a national scale, is likely to cause the decline of native species. To*

determine the spatial distribution of the species, the federation decides to organize presence records at various strategic points of the river network. From these surveys, the federation wants to characterize the river sections concerned by the presence of this species. An analysis of simple environmental variables (Strahler order, flow velocity) of all rivers in the catchment area will make it possible to identify the places likely to be colonized. The aim of the federation is to put in place management actions to limit the species spreading. For this purpose, it has a RHT on which a set of attributes are calculated.

To carry out this project, the steps are as follows:

1) verification of the network consistency: before any network analysis, it is necessary to make sure of the network consistency in terms of geometry, topology and connectivity;

2) route organization: the objective is to organize the teams' routes between observation stations;

3) alignment of points on a network: this step consists of aligning the observation points on the river network arcs;

4) network classification: this step consists of calculating the Stahler number of rivers affected by the presence of Louisiana crayfish;

5) stations characterization: calculation of the average width and the flow velocity on each network arc;

6) distance calculation between observation points: the objective is to calculate the shortest distance between observation stations;

7) upstream path and drainage basins calculation: the network upstream path, from a network arc, allows the delimitation of the corresponding catchment area;

8) downstream path: this step presents a Python script to carry out a downstream path from a network arc to the basin outlet;

9) calculation of availability areas: with locations where the presence of Louisiana crayfish is known, it is possible to estimate the spread during the movement phase.

4.4.2. *Useful data*

The thematic analysis is based on vector data in Lambert93 / RGF93 projection:

– rht_true.shp: theoretical river network (polyline);

– rht_false.shp: river network with geometry and topology errors;

– sta.shp: observation stations (point);

– surfeau.shp: water surface (polygon);

– fr50bv.shp: subdrainage basin (polygon).

4.4.3. *Step 1 – verification of network consistency*

The first step is to ensure that the network is coherent in terms of geometry, topology and connectivity. This section presents the checks necessary before any network analysis. It also indicates how to correct different types of errors.

4.4.3.1. *Geometry checking*

Step	QGIS handling
1. Geometry Checker plugin installation	In **QGIS**: • Click on **Plugins > Manage and Install Plugins...** • In the **Plugins** window, search Geometry Checker • Check the tool (normally installed by default).
2. Tool description	The Geometry Checker is one of the main QGIS plugins. It enables the user to check and repair the geometry of a layer. For linear vectors, this tool checks the following errors: • self-intersection: a segment that intersects itself; • duplicate nodes: two duplicate nodes in a segment; • segment length: a segment length smaller than a threshold; • minimal angle: two segments with an angle smaller than a threshold.
3. False file opening	• Open the following file: o rht_false.shp: river network with geometry errors
4. Geometry errors detection	In the **menu bar**: • Click on **Vector > Geometry Tools > Check Geometries** In the **Check Geometries** window: • Select the input vector file: rht_false.shp • In the **Geometry validity** field, check the options: o Self intersections o Duplicate nodes • In the **Allowed geometry types** field, check the option: o Line • Save the output vector layer as: rht_correction.shp • Click on **Run** In the **Check Geometries** window, the **Result** tab is activated and presents the detected geometry errors.

<table>
<tr>
<td></td>
<td>In the Check Geometries window,
• Click on the button Export
• Save the layer as: rht_false_errors.shp
The created vector file contains a point layer that identifies errors and their type.</td>
</tr>
<tr>
<td>5. Detected errors fix</td>
<td>In the Check Geometries window, in the Result tab:
• Select the error corresponding to the object ID: 345
The map canvas is then centered on the error, here a self-intersection:
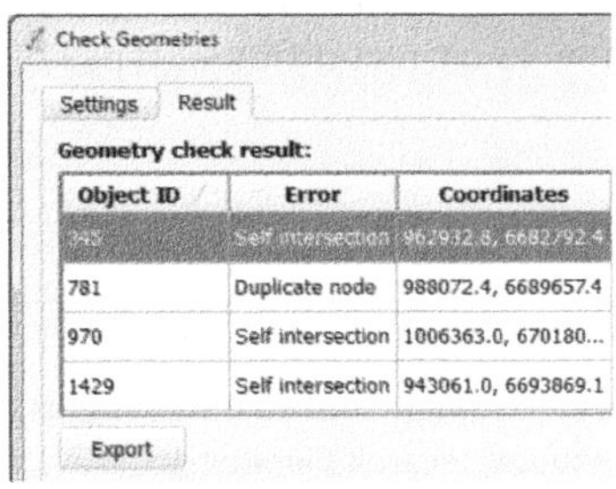

• Click on the button:
o Fix selected errors, prompt for resolution method
• Select the option:
o Split feature into multiple single-object features
• Click on the button Fix.
The rht_correction.shp file is updated. The false feature is split into several separate features. It is then up to the user to manually delete the erroneous elements.
• Fix the remaining errors by choosing the appropriate options.
• Once the errors have been fixed, click on the Close button.
• Using the rht_false_errors.shp layer,
o Find the self-intersection errors on the rht_correction.shp layer
o Delete unnecessary features (by enabling Edit mode).</td>
</tr>
</table>

Table 4.1. *Geometry checking. For a color version of the table, see www.iste.co.uk/baghdadi/qgis4.zip*

4.4.3.2. *Topology checking*

Step	QGIS handling
1. Topology Checker plugin installation	In the **menu bar**: • Click on **Plugins > Manage and Install Plugins…** In the **Plugins** window, search Topology Checker and, • Check the tool (normally installed by default).
2. Topology concepts in GIS	The topology checker allows analysis of vector layers by testing different spatial relation rules between connected or adjacent vector features. There are different types of topological error that can be grouped according to the vector type (point, line or polygon). For example, with regard to linear features, the most common errors are: – self-intersection, if lines intersect or superimpose themselves;  – overlap, if lines overlap with other lines of the same layer;  – dangles, if the end of a line is not connected to at least one other line;  – pseudo-nodes, if a line is not connected to at least two other lines at each end.  Topological errors break the connection between vectors and must be fixed in order to be able to process data in network analysis.
3. Vector file opening	• Open the previously created file: rht_correction.shp: river network with geometry fix

4. Search for duplicates	In the **menu bar**: • Click on **Vector** > **Topology Checker** > **Topology Checker** • Or click on . In the **Topology Checker Panel**, • Click on the **Configure** icon In the **Topology Rule Settings** window, • Enter the following rule: rht_correction must not have duplicates o Click on the **Add Rule** button o Click on **OK** In the **Topology Checker Panel**, • Click on the **Validate all** icon • Edit the rht_correction layer • Delete duplicates • Restart the test.
5. Search for dangles	• In the same way, check the following rule: rht_correction must not have dangles N.B. In the case of river networks, this topological rule is not always respected. Indeed, nodes corresponding to sources and outlets are not linked to other lines. In the present case, no correction is to be expected.
6. Search for pseudo-nodes	• Check the following rules: rht_correction must not have pseudos In the present case, no correction is to be expected.

Table 4.2. *Topology checking. For a color version of the table, see www.iste.co.uk/baghdadi/qgis4.zip*

4.4.3.3. *Connections checking*

<table>
<tr><th>Step</th><th>QGIS handling</th></tr>
<tr><td>1. Disconnected Islands plugin installation</td><td>In the menu bar:
• Click on Plugins > Manage and Install Plugins…
In the Plugins window,
• Search Disconnected Islands
• Click on Install plugin.</td></tr>
<tr><td>2. Tool description</td><td>This tool is used to check the connectivity of an undirected network (or graph). It assigns to each subgraph (or connected component) an identifier. It makes it possible to restore the missing links if necessary.
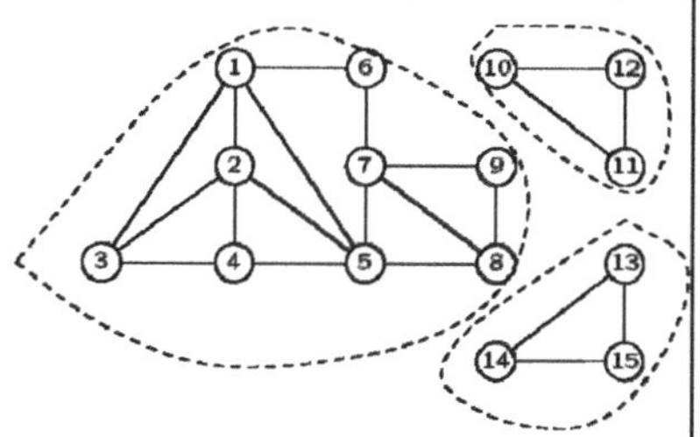
</td></tr>
<tr><td>3. Vector file opening</td><td>• Open the previously created file:
  o rht_correction.shp: river network with geometry and topology fix.</td></tr>
<tr><td>4. Subgraphs identification</td><td>In the menu bar:
• Click on Vector > Disconnected Islands > Check for Disconnected Islands
• Or click on .
In the Disconnected Islands window,
• Select the rht_correction layer
• Modify the tolerance (if topology is not perfect)
• Enter an attribute name for storing subgraphs identifiers, for example:
  o Attribute name: id_graph
• Click on OK.</td></tr>
<tr><td>5. Analysis</td><td>The river network actually has eight subgraphs. Some of them are unnecessary for this study.</td></tr>
</table>

Subgraphs consisting of a single tributary (a single vector feature) are considered here as errors (id_graph 4 to 8). They should be deleted.

In the **Layers Panel**:

- Select the rht_correction layer
- Open the **Field Calculator** by clicking on

In the **Field Calculator**:

- Check the option **Create a new field**
- Set as follows:
  - **Output field name**: Nb_Lines
  - **Output field type**: whole number (integer)
- Enter the following expression:

```
count( "id_graph" , "id_graph" )
```

- Click on **OK**
- Save modifications
- Click on

In the **Select by expression** window,

- Enter the following expression:

```
"Nb_Lines" = 1
```

- Click on **Select** button
- Delete selected features.

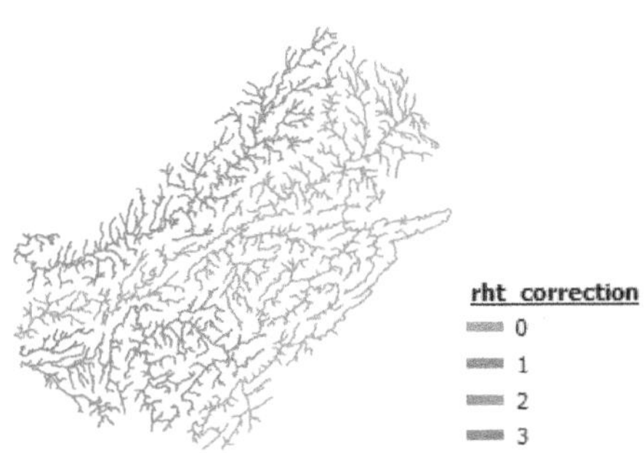

Subgraphs of RHT network

Table 4.3. *Subgraphs detection. For a color version of the table, see www.iste.co.uk/baghdadi/qgis4.zip*

4.4.4. *Step 2: routes organization*

In GIS, network analysis is widely used to calculate the shortest route between two points. As part of our study, we are considering a field campaign to collect measurements at different observation points. The objective is to organize the teams' routes between stations no. 24 and no. 15, no. 2 and no. 14, no. 3 and no. 12.

Step	QGIS handling
1. Vector files opening	• Open the following files: o rht_true.shp: theoretical river network o sta.shp: observation stations
2. Online Routing Mapper plugin installation	In the **menu bar**: • Click on **Plugins > Manage and Install Plugins…** In the **Plugins** window, • Search Online Routing Mapper • Click on **Install plugin** • Launch the plugin by clicking on: o **Plugins > Online Routing Mapper > Online Routing Mapper** o Or click on .
3. Tool description	The application calculates the shortest path between two points. However, the result may be very different depending on the data used. For example, the analysis will be biased if the roads do not contain directional information (unidirectional or bidirectional). Map services such as Google Maps, BingMap, etc. allow you to not only be positioned in relation to one place (coordinates, toponym and address) but also to calculate routes from one city to another one. The application uses these services to calculate routes without having the data on the workstation. The results can be exported in different formats, including the Shapefile.

<table>
<tr><td></td><td>1)
2)
3)
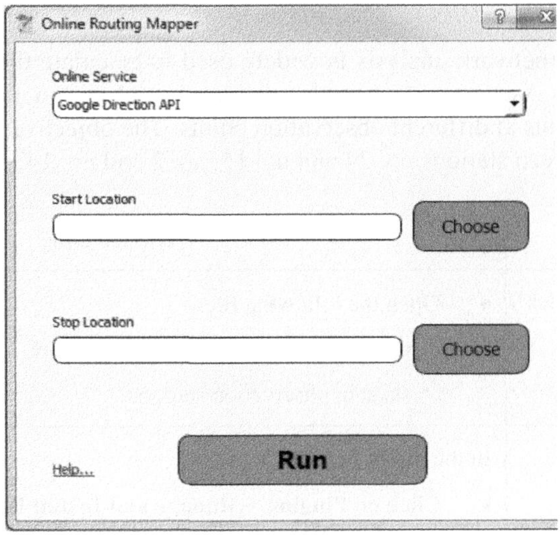

1. Map service choice,
2. Starting point coordinates,
3. Finish point coordinates.</td></tr>
<tr><td>4. Routes computation</td><td>In the Online Routing Mapper window:
• Set as follows:
o Online Service: Google Direction API
o Start Location:
▪ Click on the green Choose button
▪ Click on station no. 24
o Stop Location:
▪ Click on red Choose button
▪ Click on station no. 15
• Click on Run button
• In the same way, compute the routes between the following stations:
o no. 2 and no. 14
o no. 3 and no. 12.</td></tr>
</table>

	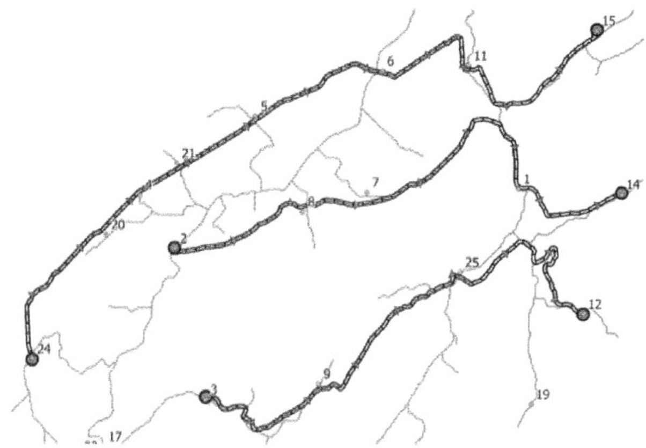 Routes between observation stations.

Table 4.4. *Calculation of road itineraries. For a color version of the table, see www.iste.co.uk/baghdadi/qgis4.zip*

4.4.5. *Step 3: alignment of points on a network*

This step consists of aligning (snap) the observation points on the river network's arcs. It makes it possible to generalize the presence information at the level of the corresponding arc on the one hand, and to use environmental attributes previously calculated on the network on the other hand. This processing is done using a Python script.

Step	QGIS handling
1. Vector file opening	• Open the following files: o rht_true.shp: theoretical river network o sta.shp: observation stations • Check in the **Layers Panel** that the layers order displays as follows: 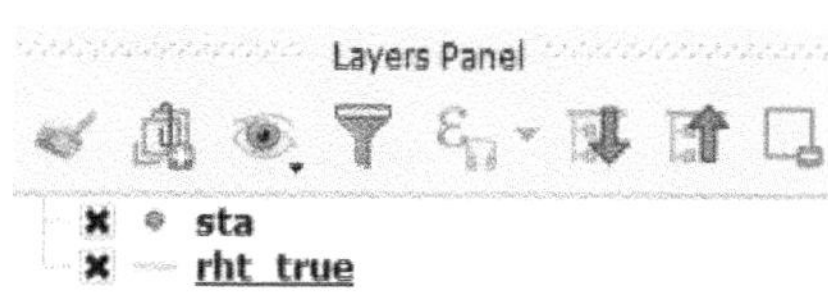

2. Discovery of Python console	In the **menu bar**: • Click on **Plugins > Python Console**. The Python Console is displayed at the bottom of the screen: 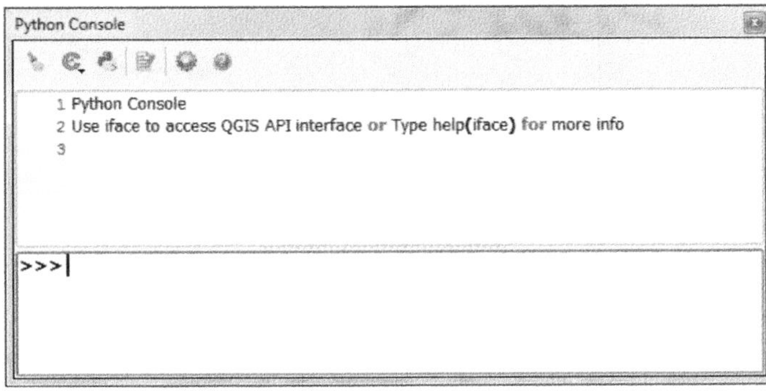 In the **Python Console**: • Click on the **Show Editor** icon The script editor allows you to launch a series of instructions in order to carry out sometimes complex treatments. When a script is run, the console communicates to the developer information, results and errors.
3. Script opening	In the **Python Console**: • Click on the **Open Script** icon • Open the script: snap_points.py
4. Script operating	The script seeks, for each point of the sta.shp layer, the closest arc in the rht_true.shp layer. It creates at the output a temporary vector layer containing the points aligned on the river network and calculates the distance that separates the origin point and the arrival point.
5. Script running	In the **Python Console**: • Click on the **Run script** icon The script creates a new temporary layer of points, ptsnap_v1, aligned on the river network. In the dist field, the attribute table has the snapping distance.

	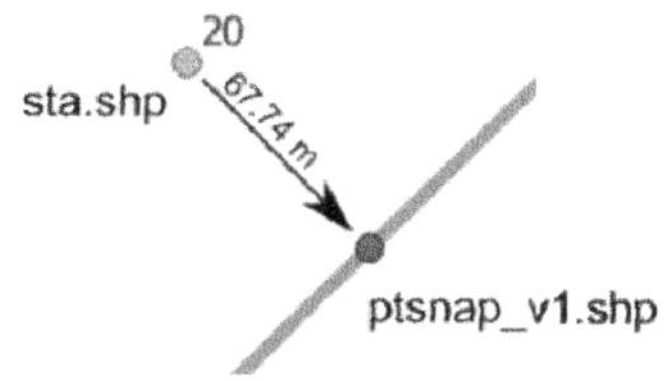 • Save the result as ptsnap_v1.shp

Table 4.5. *Alignment of observation points on the network. For a color version of the table, see www.iste.co.uk/baghdadi/qgis4.zip*

4.4.6. *Step 4: network classification*

The fishing federation wants to know the Strahler stream order of rivers affected by the presence of the Louisiana crayfish. QGIS provides a plugin dedicated to the calculation of the Strahler number.

Step	QGIS handling
1. Strahler plugin installation	In the **menu bar**: • Click on **Plugins** > **Manage and Install Plugins...** In the **Plugins** window, • Search Strahler • Click on **Install plugin**.
2. Tool description	The Strahler plugin allows the Strahler number to be calculated based only on the network topology (without a digital terrain model). The network must be divided into segments. For example, a "T" junction must consist of three segments (not two). Similarly, an "X" junction has four segments, etc. The plugin calculates the Strahler number for each segment and adds this information in a new field of the attribute table. N.B. This tool does not support polyline features.
3. Vector file opening	• Open the following file: o rht_true.shp: river network
4. Example	• Select the segment on the network of rht_true.shp layer, which represents the outlet (field "GID" = 3306). This segment must be connected to the rest of the network by a single node.

	• Run the plugin by clicking on **Vector > Strahler > Strahler** • Or click on . In the **Strahler** window, • Enter a **new attribute name** which stores the Strahler number, for example: Strahler • Click on **OK** • Open the attribute table • Check that the new field is created and filled in.  Strahler stream order.

Table 4.6. *Calculation of Strahler order. For a color version of the table, see www.iste.co.uk/baghdadi/qgis4.zip*

4.4.7. *Step 5: stations characterization*

The movement and survival of Louisiana crayfish are assumed to depend on average flow velocities. This variable, based on the averages of the waterway width, height and flow, is calculated as follows:

$V = Q/S$,
where V is the average speed in m/s^{-1}, Q is the average flow in m^3/s^{-1} and S is the flow surface in m^2.
With:
$S = H \times L$,
where H is the average height in meters and L is the average width in meters.

All these parameters are previously computed on the RHT. First, we present a method for estimating the average width of some sections as a function of a water surface's vector layer (surfeau.shp). Then, we will compute the flow velocities on the observation points (ptsnap_v1.shp), taking into account the presence and absence of data of the studied species.

4.4.7.1. *Estimation of the average width of water surfaces*

The strategy for calculating the average width of the water surface consists of creating perpendicular (cross-section) and regularly spaced (here 5 m) segments along the river (long profile). The segments length is then averaged. This estimate will be performed at three locations in the network using the surfeau.shp vector layer.

Step	QGIS handling
1. Vector files openings	• Open the following files: o rht_true: river network o surfeau.shp: water surface o ptsnap_v1.shp: stations aligned on the network
2. Spatial Query plugin installation	In the **menu bar**: • Click on **Plugins > Manage and Install Plugins…** In the **Plugins** window: • Search Spatial Query Plugin • Check the tool.
3. Segment selection	In the **menu bar**: • Click on **Vector > Spatial Query > Spatial Query** • Or click on  In the **Spatial Query** window: • Select source features from: rht_true • Choose the spatial rule: Intersects • Select the reference features of surfeau • Use the result to **Create a new selection** • Click on the **Apply** button
4. Station Lines plugin installation	In the **menu bar**: • Click on **Plugins > Manage and Install Plugins…** In the **Plugins** window: • Search Station Lines • Install the tool.

5. Creation of cross-sections	In the **menu bar**:
	• Click on **Plugins** > **Station Lines** > **Station Lines**
	• Or click on 
	In the **Station Lines** window:
	• Select the rht_true layer
	• Check the option **Use only selected features**
	• In the **Stations choices** box:
	  o Check the parameter Distance
	  o Enter the value of 5 (meters)
	• In the **Side** box:
	  o Choose the value of Both
	• In the **Length** box:
	  o Enter a value of 500 (meters)
	• In the **Angle** box:
	  o Enter a value of 90 (degrees)
	• Click on **OK**
	A temporary vector layer, named Station Lines, appears on the screen. It has six attributes:
	• FID: network feature ID,
	• SL_ID: cross-section ID,
	• SL_SEGMENT: linear feature segment ID,
	• SL_ANGLE: cross-section angle,
	• SL_LENGTH: cross-section length,
	• SL_ORIENT: cross-section direction.

6. Intersection of cross-sections with water surface	In the **menu bar**: • Click on **Processing** > **Toolbox** In the **Processing Toolbox**: • Double-click on: o **QGIS geoalgorithms** > **Vector overlay tools** > **Intersection** In the **Intersection** window, set as follows: • **Input layer**: Station Lines • **Intersect layer**: surfeau • Create **temporary layer** • Click on **Run** A new temporary layer, named Intersection, appears on the screen.

<table>
<tr><td></td><td>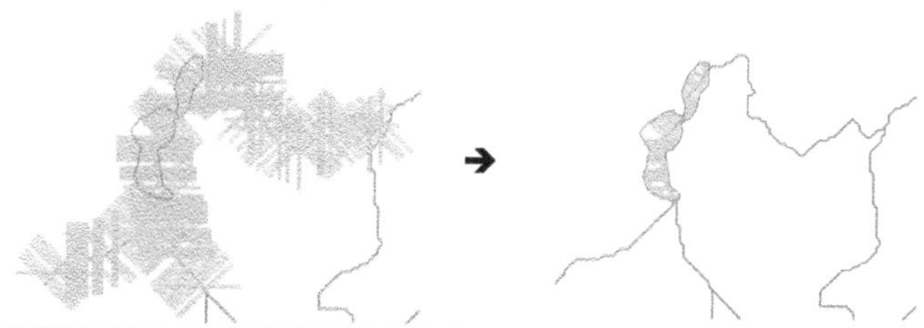</td></tr>
<tr><td>7. Cross-sections length update</td><td>
In the Layers Panel:
<ul>
<li>Select the Intersection layer</li>
<li>Open the Field Calculator by clicking on</li>
</ul>
In the Field Calculator:
<ul>
<li>Update the SL_LENGTH field</li>
<li>Enter the following expression:</li>
</ul>
$length
<ul>
<li>Click on OK</li>
<li>Save edits</li>
</ul>
</td></tr>
<tr><td>8. Calculation of water surfaces average width</td><td>
In the Processing Toolbox:
<ul>
<li>Double-click on:
<ul><li>QGIS geoalgorithms > Vector table tools > Statistics by categories</li></ul>
</li>
</ul>
In the Statistics by categories, set as follows:
<ul>
<li>Input vector layer: Intersection</li>
<li>Field to calculate statistics on: SL_LENGTH</li>
<li>Field with categories: TOPONYME</li>
<li>Save to temporary file</li>
<li>Click on Run</li>
</ul>
The output result is a table showing the three studied water surfaces in rows, the statistics relating to the SL_LENGTH field in columns. The mean field is the average width.
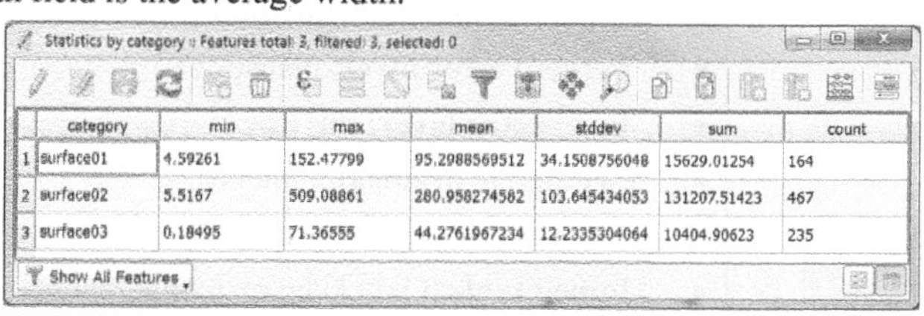
Statistics by category :: Features total: 3, filtered: 3, selected: 0
<table>
<tr><th></th><th>category</th><th>min</th><th>max</th><th>mean</th><th>stddev</th><th>sum</th><th>count</th></tr>
<tr><td>1</td><td>surface01</td><td>4.59261</td><td>152.47799</td><td>95.2988569512</td><td>34.1508756048</td><td>15629.01254</td><td>164</td></tr>
<tr><td>2</td><td>surface02</td><td>5.5167</td><td>509.08861</td><td>280.958274582</td><td>103.645434053</td><td>131207.51423</td><td>467</td></tr>
<tr><td>3</td><td>surface03</td><td>0.18495</td><td>71.36555</td><td>44.2761967234</td><td>12.2335304064</td><td>10404.90623</td><td>235</td></tr>
</table>
Show All Features
</td></tr>
</table>

Table 4.7. *Estimation of the average width of water surfaces. For a color version of the table, see www.iste.co.uk/baghdadi/qgis4.zip*

4.4.7.2. *Estimation of flow speed*

The wetted width is a parameter that can be calculated by a hydraulic geometry model. This model estimates, for each network section, values of water height and wetted width [MIG 16] as a function of the modeled flow rate [SAU 06]. This estimate will be used to calculate the flow speed.

Step	QGIS handling
1. Precomputed variables	The theoretical river network has useful variables for calculating the flow speed: • mod: average flow rate ($m^3 \cdot s^{-1}$) • haut: average height (m) • larg: average width (m)

2. Flow speed calculation	In the **Layers Panel**:
	• Select the rht_true layer
	• Open the **Field Calculator** by clicking on
	In the **Field Calculator**:
	• Create a new field as follows:
	  o **Output field name:** speed
	  o **Output field type:** Decimal number (real)
	  o **Ouput field length:** 10, **Precision:** 2
	• Enter the following expression:
	`"mod" / ( "haut" * "larg" )`
	• Click on **OK**
	• Save edits

3. Spatial join	In the **Processing Toolbox**: • Double-click on: o **QGIS geoalgorithms > Vector general tools > Join attributes by location** In the **Join attributes by location** window, set as follows: • **Target vector layer**: ptsnap_v1 • **Join vector layer**: rht_true • **Geometric predicate**: intersects • **Attribute summary**: Take attributes of the first located feature • **Joined table**: Only keep matching records • **Joined layer**: ptsnap_v2.shp • Click on **Run** The new ptsnap_v2.shp layer appears on the screen.

4. Analysis	By analyzing the attribute table of ptsnap_v2.shp layer, we roughly deduce that the Louisiana crayfish is constrained to areas where flow velocity is lower than 0.5 m/s^{-1}. 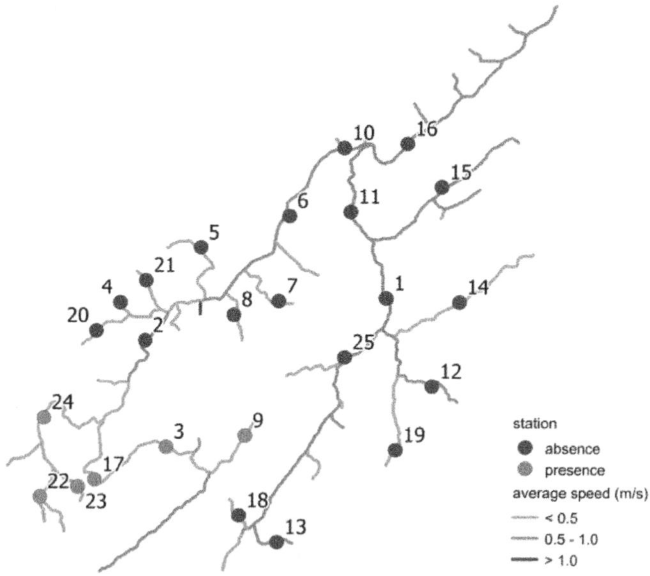

Table 4.8. *Estimation of flow speed. For a color version of the table, see www.iste.co.uk/baghdadi/qgis4.zip*

4.4.8. *Step 6: distance calculation between observation points*

This section presents the distance calculation via a network. QGIS provides a plugin to extract the shortest distance between two points. It is also possible to systematize the calculation over several pairs of points using the QGIS API with a Python script.

4.4.8.1. *Distance calculation between two points*

The objective here is to calculate the shortest distance between observation stations no. 24 and no. 25 via the RHT.

Step	QGIS handling
1. Road graph plugin installation	In the **menu bar**: • Click on **Plugins** > **Manage and Install Plugins…** In the **Plugins** window: • Search Road graph plugin • Check the tool.

2. Vector files opening	• Open the two following files:
	  o rht_true.shp: river network
	  o sta.shp: stations

3. Plugin setup	In order to work, the plugin must be configured beforehand.
	In the **menu bar**:
	• Click on **Vector** > **Road Graph** > **Settings…**
	The **Settings** window is displayed. It allows you to configure the transport layer, the graph direction and the type of weight taken into account in the calculation (length or duration).
	In the **Settings** window:
	• In the **Transportation layer** tab:
	  o Select the rht_true layer
	• In the **Default settings** tab:
	  o Select a Two-way direction
	• Click on **OK**

4. Calculation of the shortest path between two stations	In the **Road Graph** window (called **Shortest path**), normally placed to the left under the Layers Panel: • Enter the start point coordinates by first clicking on the icon and then on the station no. 24 displayed in the map canvas • In the same way, enter the end point coordinates (station no. 25) • Select the Length criterion • Click on the **Calculate** button The shortest path between the two selected stations is highlighted. The length that separates the two points is approximately 39 km.

	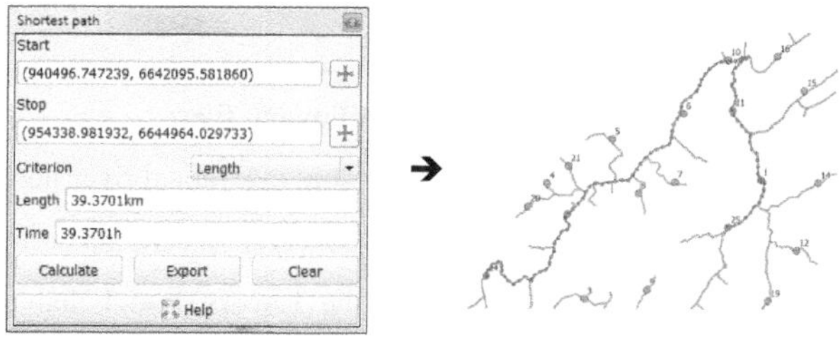 N.B. The tool can weigh the shortest path according to the travel time. This criterion is widely used in road networks where sections have different speed limits. In hydrology, as graphs are usually hierarchical, the geometry of the shortest path is not modified, whatever the chosen criterion. It is possible to calculate a transfer speed of water mass from one point to another on the network using the average flow velocity per arc.
5. Result recording	In the **Road Graph** window: • Click on the **Export** button In the **Export feature** window: • Select a New temporary layer • Click on OK A temporary layer, named shortest path, appears in the Layers Panel. In the **Layers Panel**: • Right-click on the shortest path layer • Select **Save as**… • Save the layer as path_sta_24_25.shp

Table 4.9. *Distance calculation between two observation points. For a color version of the table, see www.iste.co.uk/baghdadi/qgis4.zip*

4.4.8.2. *Systematic distance calculation*

The systematic distance calculation between stations via the network can be automated using the QGIS API. The Python script here lists all the possible pairs of stations and then calculates the shortest distance for each.

Step	QGIS handling
1. Vector file opening	• Open the following files: o rht_true.shp: theoretical river network o sta.shp: observation stations • Check in the **Layers Panel** that the layers order displays as follows: 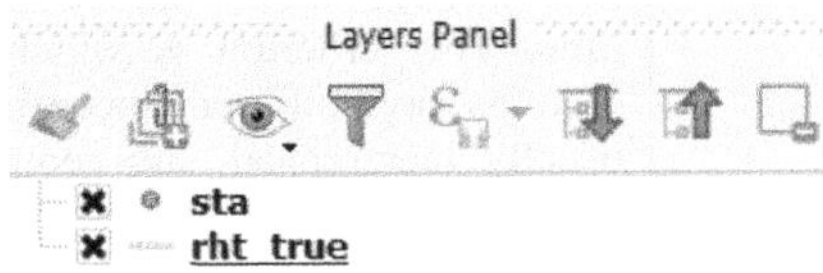
2. Script opening	In the **menu bar**: • Click on **Plugins > Python Console**. In the **Python Console**: • Click on the **Show editor** icon In the **Python Console**: • Click on the **Open script** icon • Open the script: calc_distance_fct.py
3. Script operating	In a first step, the script constructs in memory all possible associations between stations. An association contains the two stations' identifiers and their coordinates. Then, using the network analysis module provided by the QGIS API, it determines the shortest paths via the river network. Calculated distances are displayed in the console while the route is highlighted in red in the map canvas.
4. Script running	In the **Python Console**: • Click on the **Run script** icon

	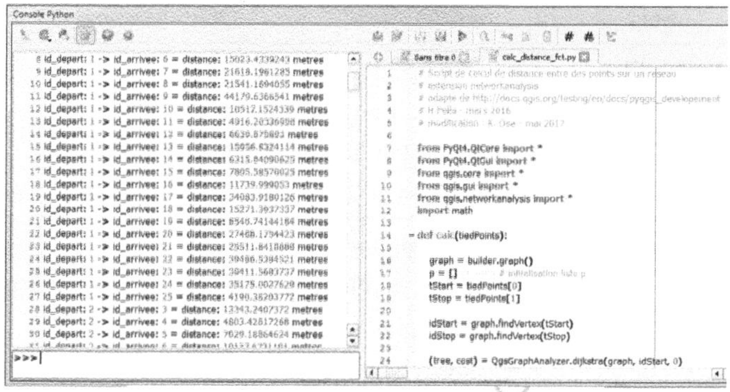 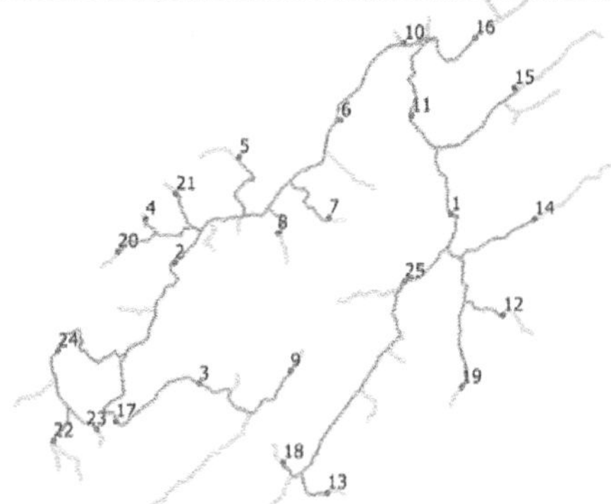 The Python script displays in the console the shortest distances for each pair of stations along the river network. The shortest paths appear highlighted in red in the software map canvas.

Table 4.10. *Distance calculation between all observation points. For a color version of the table, see www.iste.co.uk/baghdadi/qgis4.zip*

4.4.9. *Step 7: upstream path and drainage basins calculation*

QGIS provides a plugin to cross the network upstream from a selected arc. This result allows the identification by intersection of subdrainage basins which, once merged, delimit the overall supply area. From drainage basins associated with each of the observation points, it is possible to deduce and analyze surface type variables (land cover, discharge, etc.).

Step	QGIS handling
1. Vector files opening	• Open the two following files: o rht_true.shp: river network o fr50bv.shp: drainage basins o ptsnap_v1.shp: stations aligned on the network
2. Flow trace plugin installation	In the **menu bar**: • Click on **Plugins** > **Manage and Install Plugins...** In the **Plugins** window: • Search Flow Trace • Install the tool.
3. Upstream path from a selected arc	The objective is to cross upstream from observation station no. 6 and to deduce the limits of the corresponding catchment area. • Select the rht_true network's arc ("GID" = 3764) which crosses station no. 6 in the ptsnap_v1 layer • Click on the **Flow Trace** plugin icon . The Flow Trace tool selects all arcs located upstream of the selected arc.
4. Drainage basins selection	In the **menu bar**: • Click on **Vector** > **Spatial Query** > **Spatial Query** • Or click on In the **Spatial Query** window: • Select source features from: fr50bv • Where the feature: Intersects • Reference features of: rht_true • Check the **Selected geometries** option • Use the result to: Create new selection • Click on **Apply** In the **menu bar**: • Click on **Processing** > **Toolbox** In the **Processing Toolbox**: • Double-click on: o **QGIS geoalgorithms** > **Vector geometry tools** > **Dissolve** In the **Dissolve** window, set as follows: • **Input layer**: fr50bv • Check the **Dissolve all** option

	• Create a **temporary layer** • Click on **Run** A new temporary layer, named Dissolved, appears on the screen.
	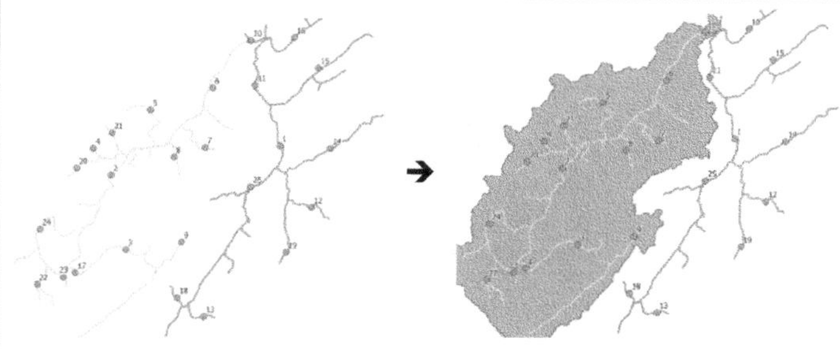 Upstream path Drainage basin

Table 4.11. *Upstream path. For a color version of the table, see www.iste.co.uk/baghdadi/qgis4.zip*

4.4.10. *Step 8: downstream path*

Despite the QGIS plugin for crossing networks upstream, there is currently no plugin to follow the flow downstream. One of the possible alternatives is to develop a Python script using the QGIS API with the network analysis library module. Besides describing the algorithm, the goal is also to become familiar with writing scripts to run processes that do not have an interface.

Step	QGIS handling
1. Vector file opening	• Open the following file: o rht_true.shp: river network • Select the rht_true network's arc ("GID" = 3764) which crosses station no. 6 in the ptsnap_v1 layer
2. Script opening	In the **menu bar**: • Click on **Plugins > Python Console** In the **Python Console**: • Click on the **Show editor** icon In the **Python Console**: • Click on the **Open script** icon • Open the script: parcoursAval.py

<table>
<tr><td>3. Script description</td><td>
To fully understand how this tool works and the Python functions used, the script is described in detail. This code largely follows the instructions provided in the QGIS Developer Cookbook, available on the Web[2].
a) Script header
Before coding, it is strongly advised to add in the first lines some information about the script's objectives, its author, its creation date or its last update. Comments are preceded by the # symbol. At run time, these lines are not taken into account. More generally, do not hesitate to comment on each step of treatment. A good script is a commented script!
<pre>
Script objectives: ...
QGIS version <= 2.16
Author: ...
Date of creation: ...
</pre>
b) QGIS module import for network analysis
By default, a Python script can call a number of basic functions. For network analysis, it is imperative to import the dedicated tools. These are provided by the Network Analysis module of the QGIS API.
<pre>
# Network analysis module import
from qgis.networkanalysis import *
</pre>
This instruction can be read as follows: "In the QGIS API, we import all the tools proposed by the Network Analysis module".
c) Loading river network layer
The first step is to load the river network vector file.
<pre>
# Get the active vector layer in Layers Panel
layer = iface.activeLayer()
</pre>
At the same time, we are looking for the projection system of this layer, as well as the first selected arc.
<pre>
# Get the CRS (projection)
crs = layer.crs()
# Get the first selected feature
plsel = layer.selectedFeatures()[0]
</pre>
d) Directed graph building
In a second step, we transform the vector layer into a directed graph. All subsequent actions will be run on this graph and not on the vector.
</td></tr>
</table>

2 http://docs.qgis.org/testing/en/docs/pyqgis_developer_cookbook/networkanalysis.html.

In QGIS, the network, converted into a graph, is directed by the QGSLineVectorLayerDirector function. It takes as input six parameters:

– vl: vector layer used to build the graph;

– directionFieldId: index of the attribute table field where information about direction is stored (–1 if no information);

– directDirectionValue: field value for direct direction (moving from first line point to last one);

– reverseDirectionValue: field value for reverse direction (moving from last line point to first one);

– bothDirectionValue: field value for bidirection (moving in both directions);

– defaultDirection: default direction if direction field values are not informed (1 = direct direction, 2 = reverse direction, 3 = both directions).

The rht_true layer does not have a field with line directions. However, this information is not necessary since the digitization direction is in accordance with the flow direction. The objective is to go downstream of the network, so we choose a direct direction.

```
# Building directed graph
director = QgsLineVectorLayerDirector(layer, -1, '', '', '', 1)
```

It is then necessary to create a strategy for calculating edge properties, in particular the lengths of tributaries. This strategy is added to the directed graph.

```
# calculation of edges properties
properter = QgsDistanceArcProperter()
director.addProperter(properter)
```

Now we can create a Graph object using a class constructor (builder). It can take several arguments, in particular the coordinate reference system, the topological tolerance, etc.

Here we specify only the projection system (crs); the other parameters are set by default.

```
# Class constructor
builder = QgsGraphBuilder(crs)
```

We want to go through the network downstream of a given point. It is therefore necessary to add a vertex to the graph by indicating its coordinates (end of the selected arc).

```
# Adding a start point to the graph
pStart = plsel.geometry().asPolyline()[0]
tiedPoints = director.makeGraph(builder, [pStart])
```

	It remains to build the Graph object. It may take some time depending on the input vector complexity. Analysis will be carried out only on this graph.
	`# Building the graph`
	`graph = builder.graph()`

	e) Graph weighting The graph analysis is used to identify the connected vertices and find the shortest paths. It is based on the construction of a directed and weighted (by the edge length for instance) graph, also called "shortest paths tree". Its properties are as follows: – only one vertex has no incoming edges, the root of the tree; – all other vertices have only one incoming edge; – if vertex B is reachable from vertex A, then the path from A to B is the single available and optimal path (shortest) on this graph. To compute the shortest path tree, QGIS API provides two methods, shortestTree() and dijkstra(), associated with the QgsGraphAnalyzer class. Here, we test the shortestTree() method that accepts three variables: – source: input graph (here the graph variable) – startVertexIdx: index of the start point on the tree (the root of the tree) – criterionNum: number of edge properties to use (started from 0).

	`# Building of shortest paths tree`
	`startId = graph.findVertex(tiedPoints[0])`
	`tree = QgsGraphAnalyzer.shortestTree(graph, startId, 0)`

	f) Iteration on graph To read the graph, it is necessary to retrieve the edges' coordinates (start vertex and end vertex), to store them in a new vector layer (vl). We first create the temporary vector layer:

	`# Creation of output vector (temporary file)`
	`vl = QgsVectorLayer("LineString?crs=" +crs.toWkt(), "temp_line", "memory")`
	`pr = vl.dataProvider()`

	We then cross the graph edges using a loop:

	`# Graph reading`
	`i = 0`
	`while (i < tree.arcCount()):`
	    `fet = QgsFeature()`
	    `# Start point`

	lineStart = tree.vertex(tree.arc(i).inVertex()).point()
	# End point
	lineEnd = tree.vertex(tree.arc(i).outVertex()).point()
	# Creation of geometry
	fet.setGeometry(QgsGeometry.fromPolyline([lineStart, lineEnd]))
	pr.addFeatures([fet])
	vl.updateExtents()
	i = i + 1

	g) Display of result The last step of the script is to display the temporary vector layer on the QGIS map canvas.

	# Display the vector file in the QGIS map canvas
	QgsMapLayerRegistry.instance().addMapLayer(vl)

4. Script running	In the **Python Console**: • Click on the **Script running** icon 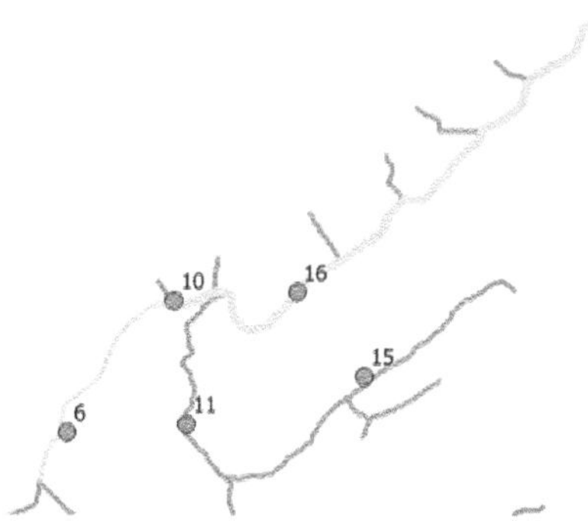 Downstream path
5. What about upstream path?	To cross the network upstream from a given point, it is sufficient to modify a parameter when creating a directed graph. The value of the defaultDirection option for the QgsLineVectorLayerDirector function is then 2 (reverse direction).

	# Building directed graph (upstream)
	director = QgsLineVectorLayerDirector(layer, -1, '', '', '', 2)

Table 4.12. *Downstream path. For a color version of the table, see www.iste.co.uk/baghdadi/qgis4.zip*

4.4.11. *Step 9 – calculation of availability areas*

Using locations where the presence of Louisiana crayfish is known, we want to calculate the dissemination during the trip phase. Crayfishes can travel for a distance of 17 km. Comparison of these availability areas with flow velocities (see section 4.4.7.2) allows us to identify network arcs potentially affected by an expansion of the species.

Step	QGIS handling
1. Introduction	The availability area of a node A is a subset of the graph accessible from the node A whose path cost is less than a given value (e.g. a length). For example: *"a crayfish can travel up to a maximum distance of 17 km along the river (upstream and downstream); which parts of the network are accessible?"* The computation of the availability area answers this question.
2. Vector files opening	• Open the following files: o rht_true: river network o sta.shp: observation stations • Check in the **Layers Panel** that the layers order displays as follows: 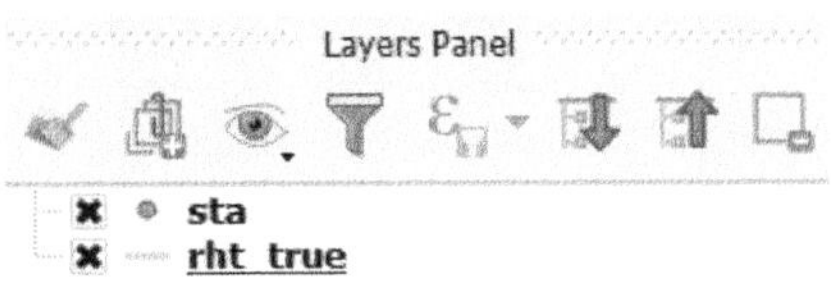
3. Processing description	The objective here is to find the network arcs potentially affected by an invasion of the Louisiana crayfish. The proposed Python script requires two input vector layers: the observation stations and the river network. The vector layer of observation stations must have two fields, named as follows: • id: number of the observation station (string); • distance: maximum distance of studied species (here 17,000 m).
4. Creation of the maximum distance field	First, we have to create the distance field and fill it in. In the **Layers Panel**: • Select the sta layer

	• Open the **Field Calculator** by clicking on
	In the **Field Calculator**:
	• Create a new field as follows:
	  o **Ouput field name**: distance
	  o **Output field type**: whole number (integer)
	  o **Output field length**: 10
	• Enter the following expression:
	17000
	• Click on **OK**
	• Save edits

5. Selection of observation stations	Second, it is necessary to select the observation stations where the presence of species is known. In the **Layers Panel**: • Select the sta layer In the **menu bar**: • Click on the **Select features using an expression** icon In the **Select by expression -sta** window, • Enter the following expression: "Presence" = 1 • Click on **Select**
6. Python script opening	The parcoursArea.py script looks in all directions (upstream and downstream) for arcs that are at a distance lower than that specified in the distance field of the sta.shp layer. It returns a temporary vector layer of accessible arcs, named temp_line. The attribute table contains two fields that indicate the number of the station concerned and the arcs' distance. In the **Python Console**: • Open the script: parcoursArea.py • Run the script
7. Speed and maximum distance parameters	The potential expansion area of the Louisiana crayfish is, for this example, determined by the average flow velocity (<0.5 m/s^{-1}) tolerated by the species and its expansion distance (17,000 m). In the **Layers Panel**:

	• Select the rht_true layer In the **menu bar**: • Click on the **Select features using an expression** icon In the **Select by expression –rht_true** window, • Enter the following expression: "vitesse" < 0.5 • Click on **Select** In the **menu bar**: • Click on **Vector** > **Spatial Query** > **Spatial Query** • Or click on In the **Spatial Query** window: • Select source features from: temp_line • Where the feature: Intersects • Reference features of rht_true • Check the **Selected geometries** option • Use the result to **Create new selection** • Click on **Apply** • Save the selected features of temp_line layer in a new file named: expansion_area.shp  Distance < 17 km Speed < 0.5 m/s^{-1} Intersection
8. Delimitation of the real expansion area	The result shows a network split into subgraphs. Theoretically, the crayfishes cannot pass from one network subgraph to another, since the latter are separated by hydrographic sections with flow velocities greater than 0.5 m/s^{-1}. It is therefore necessary to identify these various connected components and to retain only the one occupied by the invasive species. In the **menu bar**: • Click on **Vector** > **Disconnected Islands** > **Check for Disconnected Islands**

- Or click on .

In the **Disconnected Islands** window,

- Select the expansion_area layer
- Enter an attribute name for storing subgraphs identifiers, for example:
  - **Attribute name**: id_graphe
- Click on **OK**.

In the **Processing Toolbox**,

- Double-click on:
  - **QGIS geoalgorithms > Vector geometry tools > Dissolve**

In the **Dissolve** window, set as follows:

- **Input layer**: expansion_area
- Uncheck the **Dissolve all** option
- **Unique ID fields**: id_graphe
- **Save the result as**: real_expansion_area.shp
- Click on **Run**

Edit the real_expansion_area.shp layer and delete the feature whose value in the id_graphe field is equal to –1 (topology errors caused by the features overlapping in the expansion_area.shp layer).

By visually comparing the stations that have observed the presence of the crayfish and the real_expansion_area network, it is found that the expansion area corresponds to the upstream portion of the drainage basin.

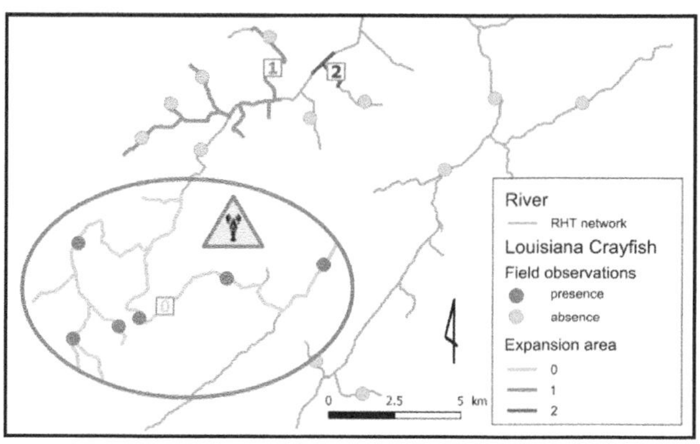

Table 4.13. *Delimitation of the real expansion area of Louisiana crayfish. For a color version of the table, see www.iste.co.uk/baghdadi/qgis4.zip*

4.5. Bibliography

[BEA 10] BEAUGUITTE L., Graphes, réseaux, réseaux sociaux: vocabulaire et notation, Version 1, Groupe f.m.r. (flux, matrices, réseaux), CNRS, UMR Géographie-cités, 2010.

[DIJ 59] DIJKSTRA E.W., "A note on two problems in connexion with graph", *Numerische Mathematik*, vol. 1, no. 1, pp. 269–271, 1959.

[DUP 13] DUPAS R., CURIE F., GASCUEL-ODOUX C. *et al.*, "Assessing N emissions in surface water at the national level: Comparison of country-wide vs. regionalized models", *Science of the Total Environment*, vol. 443, pp. 152–162, 2013.

[JOU 12] JOUANIN C., BRIAND C., BEAULATON L. *et al.*, *Eel Density Analysis (EDA2.x): a statistic model to assess European eel (Anguilla Anguilla) escapement in a river network*, Irstea Edition, available at: https://irsteadoc.irstea.fr/cemoa/PUB00036398, 2012.

[KER 13] KERR T., "The contribution of snowmelt to the rivers of the South Island, New Zealand", *Journal of Hydrology*, vol. 52, no. 2, p. 61, 2013.

[MIG 16] MIGUEL C., LAMOUROUX N., PELLA H. *et al.*, "Altération d'habitat hydraulique à l'échelle des bassins versants: impacts des prélèvements en nappe du bassin Seine-Normandie", *La Houille Blanche*, vol. 3, pp. 65–74, 2016.

[OBE 01] OBERDORFF T., PONT D., HUGUENY B. *et al.*, "A probabilistic model characterizing fish assemblages of French rivers: a framework for environmental assessment", *Freshwater Biology*, vol. 46, no. 3, pp. 399–415, 2001.

[PEL 12] PELLA H., LEJOT J., LAMOUROUX N. *et al.*, "Le réseau hydrographique théorique (RHT) français et ses attributs environnementaux", *Géomorphologie: relief, processus, environnement*, vol. 18, no. 3, pp. 317–336, 2012.

[SAU 06] SAUQUET E., "Mapping mean annual river discharges: geostatistical developments for incorporating river network dependencies", *Journal of Hydrology*, vol. 331, no. 1, pp. 300–314, 2006.

[SNE 02] SNELDER T.H., BIGGS B.J., "Multiscale river environment classification for water resources management", *JAWRA Journal of the American Water Resources Association*, vol. 38, no. 5, pp. 1225–1239, 2002.

5

Representation of the Drainage Network in Urban and Peri-urban Areas Using a 2D Polygonal Mesh Composed of Pseudo-convex Elements

5.1. Definitions and context

5.1.1. *General context and objectives*

Population growth and economic development have led to an increase in urbanization and to the transformation of rural zones into urbanized landscapes. This results in several problems related to water management [XIA 07]. Among the main problems associated with urbanization, we can highlight the increase in peak discharge [SMI 01], buildings in flooding areas, the decrease in infiltration and groundwater recharge [WMO 08], changes in base flow and hydrological regimes [MEJ 14] and the degradation of water quality due to the increase in pollution originating from urban areas [LAF 06]. To manage these territories in a sustainable way, Geographical Information System (GIS) tools able to represent realistically urban and peri-urban zones are needed. Presently GIS tools allow the delineation of catchments of medium to large scales (larger than 10 km^2), and to extract the drainage network. On the other hand, for urban and peri-urban areas, composed of a mosaic of small elements (1–100 ha), current tools do not allow a correct representation of the hydrological connectivity between the various elements of the landscape.

Chapter written by Pedro SANZANA, Sergio VILLAROEL, Isabelle BRAUD, Nancy HITSCHFELD, Jorge GIRONAS, Flora BRANGER, Fabrice RODRIGUEZ, Ximena VARGAS and Tomas GOMEZ.

There are few studies dedicated to the extraction of the geomorphological characteristics of small urban catchments [ROD 13, JAN 13]. The most common way of obtaining a spatial representation of catchments is the use of a Digital Terrain Model (DTM) [GIR 10, PAS 10, SAN 16, ZEC 94] that represents topography using a regular grid. However, these DTMs do not preserve small urban elements the size of which is often smaller than the selected DTM grid, such as some urban elements (streets, sidewalks, retention basins) or ditches. Another possibility is to use topographical representations based on triangles, Triangular Irregular Networks (TINs). Triangle meshes can capture quick changes in topography and allow the incorporation of artificial elements [BOC 12]. However, this considerably increases the number of elements in engineered zones where the topography is seldom flat and also in situations in which we want to represent small elements, characteristic of peri-urban areas. With TINs, landscape units are also lost (fields, hedges, ditches) whereas they form the base of the hydrological modeling. Another way to represent landscapes is to use irregular polygons, such as in the Geo-MHYDAS tool [LAG 10], developed for agricultural landscape, and Geo-PUMMA [SAN 17] for peri-urban areas. To use such a mesh to represent urban and peri-urban areas, a good quality mesh is required, which means elements that respect some geometrical criteria that allow a good connectivity between hydrological elements, and consequently an accurate representation of water pathways. This means elements with a gravity center inside the polygon, and convex or pseudo-convex elements. Convex elements that are very long and thin (such as roads) may also be problematic as they can act as a barrier to water flow [JAN 11]. In the following, we will assume that an irregular/heterogeneous mesh, composed of pseudo-regular elements (pseudo-convex or pseudo-squared) can be considered as a good quality mesh to extract the drainage network required for hydrological modeling. If not, we will speak of badly-shaped elements.

This chapter presents a module implemented in QGIS, belonging to the GEO-PUMMA toolbox [SAN 17]. This plugin allows a triangulation of surface elements with a vectorial format, and thus the segmentation of a mesh composed of heterogeneous polygons. This module uses the MeshPy library that is included in the Triangle® software [SHE 06]. This module allows the generation of vectorial meshes (composed of irregular polygons and linear elements) for the representation of urban and peri-urban landscapes with a high resolution, whereas classical tools do not provide a realistic solution for the determination of drainage networks in complex landscapes, highly modified by the human activity, with in particular road networks and sewer networks.

The first part describes the basic concepts needed to understand the generation of an irregular mesh in Hydrological Response Units (HRUs) [FLU 95]. Then, we present how the badly-shaped elements that affect the model mesh quality are selected. In the chain that allows the model mesh improvement, the plugin implemented in QGIS allows the decomposition of this type of polygon in triangles. We present the implementation of the TriangleQGIS and its installation in Windows, as well as the options to use it and the final result on the mesh of the Mercier catchment, a subcatchment of the Yzeron catchment, located in the suburbs of Lyon, France. This is an experimental catchment, representative of peri-urban catchments, where hydrological modeling works were conducted using the PUMMA model [FUM 17, JAN 14] and for which the required information was available at the required resolution. For one of the steps, we will also use a GIS layer from the El Guindo catchment in Chile [SAN 17].

To perform the segmentation and the generation of the drainage network, the TriangleQGIS plugin, as well as the Python scripts of Geo-PUMMA, implemented in GRASS-GIS will be used. The plugin and the Geo-PUMMA scripts are available at https://forge.irstea.fr/projects/geopumma. A Geo-PUMMA installer, as a virtual box for Windows users, is also available at the following link: http://doi.org/10.5281/zenodo.821563. The TriangleQGIS plugin can only be used in Windows and works properly until the 2.14 version of QGIS (in the virtual box, the installed version of QGIS is the 2.8). It can be used without the virtual box. On the other hand, if the user wants to realize all the steps described in the chapter to obtain the hydrological mesh, it is necessary to use Geo-PUMMA and the 6.4 version of GRASS[1], as well as the libraries that are included in the virtual box. The use of the virtual box avoids the independent installation of all the libraries with compatibility problems that can occur. Vectorial GIS layers used in the examples are located in the /home/geopumma/Database_Tutorial_GeoPUMMA_v1 folder of the virtual box and are also provided with this chapter.

The specifications of the GIS layers used in the example are the following:

Thematic content	HRUs layers of the Mercier and El Guindo subcatchment
Geographical accuracy	The minimum represented area is 10 m^2
Thematic accuracy	Associated with the input layer
Projection system	Lambert 93 – French Geodesic Network 1993 (Mercier, Lyon France) WG84/19S (El Guindo, Chili)

1 The Geo-PUMMA tool was developed with the 6.4 version of GRASS that is not compatible with version 7 of GRASS. The functions arguments used in this chapter are different between versions 6 and 7 of GRASS. Therefore, we recommend the installation of the virtual box that will avoid errors linked to the calling of GRASS 6 functions from version 7.

5.1.2. *Derivation of input GIS layers*

The first step is to gather all the GISs layers that are relevant for urban and natural areas of the catchment. In urban areas, we need a vectorial GIS layer representing the cadastral elements. It is also necessary to have the following layers: buildings, built public zones such as streets, paths, squares, parks, sport areas, regulation infrastructures and all the structures that have a hydrologically relevant role. If these data are not available, it is necessary to digitize them from maps or orthophotos. Digital Elevation Models must have a high spatial resolution, a maximum grid size of 2 m being the acceptable limit. From this information, we start delineating the catchment boundaries and its subcatchments. This task requires specific tools, as, in this kind of highly modified landscape, water pathways do not necessarily follow the topography. Jankowfsky *et al.* [JAN 13] present a methodology that integrates sewer networks and limits manual operations, combining an object-oriented approach like the one presented in this chapter for urban areas and standard DTM analyzes for rural areas. It is also necessary to collect, through field surveys, information about water flow direction in sewer networks, ditches or hydraulic structures. Once the catchment boundary is determined, all other GIS layers (land use, subcatchments, soil type, geology and others) are clipped so that all the layers have the same boundaries as the catchment ones. Subsequent steps rely on the overlay of all these layers.

Figure 5.1 shows the main GIS layers used for the generation of the hydrological mesh. These vectorial layers which are overlaid to get the input map for the algorithm are composed of polygons (land use, soil types, subcatchments, geology, etc.) and polylines (waste water and rain water sewer networks, combined sewer networks, channels, water courses, etc.). At the end of this data preparation step, land use, soil type, geology and all the other layers considered relevant for the represented hydrological processes should be available.

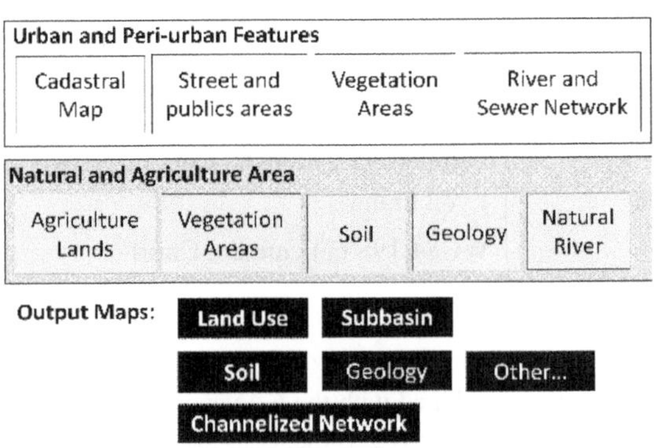

Figure 5.1. *Digitization and preparation of basic GIS layers*

All these layers must be overlaid to produce the initial vectorial mesh. Two options can be used to achieve this intersection: the first one uses the *m.seg* script, developed with the Geo-MHYDAS tool [LAG 10], which allows intersecting polygons and polylines. The second option uses the union option of QGIS applied to all layers in order to get a unique input layer. Then, it is necessary to split the polygons crossed by a polyline using the vector edition tool (split), and finally, clean the topology with the script *p.clean_topology.py* which belongs to Geo-PUMMA. This way, interfaces (limits between polygons) are preserved in the final layer and possible topology errors are corrected.

QGIS options used to join various GIS layers:
- Union tool: *Vector > Geo-treatment tools > Union...*
-Edition layer tool: *Enter the edition model then edit > Split the elements*

5.1.3. *Identification of badly-shaped HRUs and methodology to improve the model mesh quality*

The generation process of the GIS input layer includes all the relevant elements of the urban and peri-urban landscape. In the case of HRUs, polygons may be very irregular and heterogeneous. They must be segmented so that the drainage network representation is not altered. In this chapter, we do not deal with the generation of Urban Hydrological Elements (UHEs) that correspond to urban elements, composed of a cadastral field, the sidewalk and half of the adjacent road [ROD 08]. Their delineation can be performed using scripts available in Geo-PUMMA. In the following, we do not segment UHEs, the shape of which was rather regular in the treated case study. However, UHEs can also have very irregular shapes and the tools presented in this chapter may also be applied to UHEs.

Figure 5.2 shows examples of badly-shaped HRUs that were obtained after GIS vectorial layers overlay. They have an irregular shape: either non-convex or long and thin. Generally, these elements are associated with road sections or footpaths, green areas such as hedges, or the shape of agricultural fields (Figure 5.2).

These kinds of elements, that generally have their gravity center outside the polygon, lead to problems in the hydrological connectivity. This explains why it is necessary to segment them into smaller, pseudo-regular elements, without significantly increasing the final number of elements, which would affect computing performance.

Shape descriptor	Formula	Range
Form Factor, FF	$16A/P^2$	[0,1]
Convexity Index, CI	P_{convex}/P	[0,1]
Solidity	A/A_{convex}	[0,1]
Compacity	$((4/\pi)A)^{0.5}$	[0,1/π]

Table 5.1. *Considered geometrical factors. A is the polygon area, P, its perimeter and A_{convex} and P_{convex} are the area and perimeter of the convex polygons inside which the polygon is included*

A first selection of badly shaped elements is performed based on the computation of geometrical descriptors proposed by Russ [RUS 02]. Table 5.1 presents the main descriptors used that must be computed for each initial mesh. This can be achieved in QGIS, using the computation of geometrical parameters such as area and perimeter. It is also possible to get from QGIS the area of the convex polygon hull inside which the polygon is included, and then the values of the corresponding perimeter and area, P_{convex} and A_{convex}. The other option is to use the *p.shape_factors.py* Geo-PUMMA script.

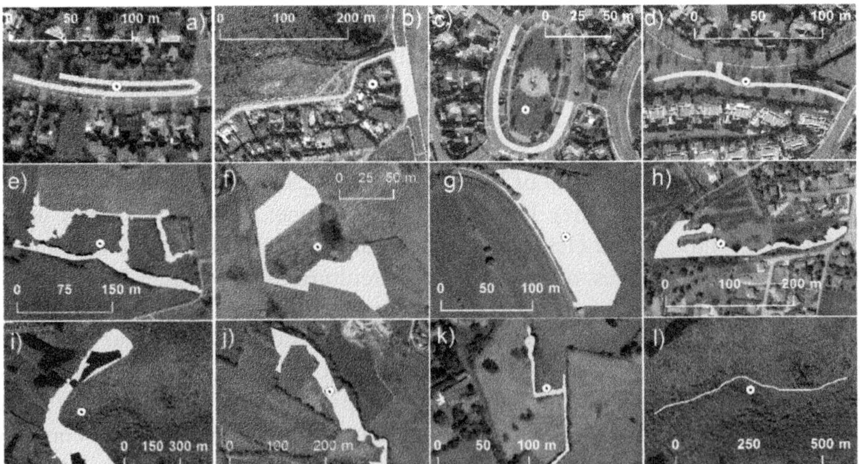

Figure 5.2. *Examples of HRUs with very irregular shapes (gray) that must be improved. These are road sections and footpaths (a, b, c, d, l), green areas (e, g, h, j, k) or agricultural fields (f, i). For a color version of the figure, see www.iste.co.uk/ baghdadi/qgis4.zip*

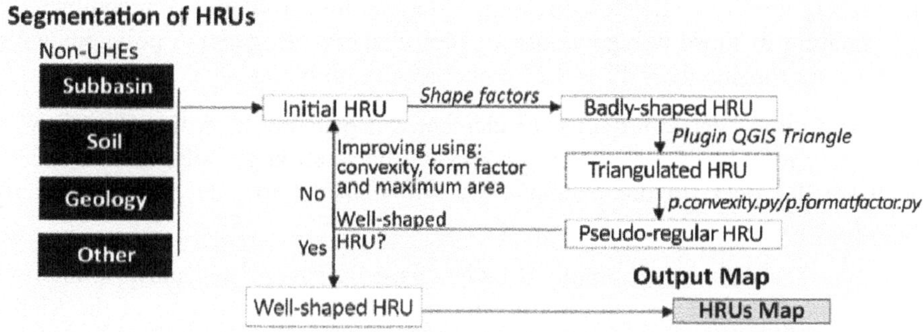

Figure 5.3. *Steps for the segmentation of badly-shaped HRUs*

> **QGIS tools for the estimation of geometrical factors:**
> - Open the calculator=> Geometry tool (this tool includes functions that compute geometries such as length, area, perimeter).

Figure 5.3 presents the HRUs generation step. In a first step, we identify badly-shaped elements from geometrical descriptors presented in Table 5.1. In a second step, the TriangleQGIS plugin is used for the triangulation of badly-shaped elements. Finally, the last step is the dissolution of triangles into pseudo-regular HRUs, according to the selected geometrical descriptor. This step allows the generation of a limited number of polygons that at best respect the initial polygon boundary, while keeping a limited number of polygons.

Criteria that allow the identification of HRUs leading to problems in the extraction of the drainage network are the following:

1) Non-convex HRUs: this type of element can generate problems in hydrological connectivity, because tools that compute the water flow distance between the polygon and its neighbor use the gravity center as the representative point of the polygons. In general, polygons with a Convexity Index IC $\geq$ 0.70 do not generate problems. In our example, we use a threshold value of the convexity index *ICT* that is equal to ICT = 0.75 to define non-convex elements.

2) Very long and thin HRUs: this kind of element can generate problems in the hydrological connectivity when an average altitude is assigned to a polygon and considered as representative. If the element is very long, this mean altitude is no longer representative and the element can act as a barrier to water flow and

erroneously modify water flux direction. To avoid this problem, we recommend the segmentation of those polygons into smaller elements. In general, polygons with a Form Factor threshold (FFT) > 0.20 do not lead to problems.

3) Too large HRUs: this kind of element can generate heterogeneities in water fluxes, in particular if the latter are computed between large polygons, adjacent to much smaller ones. In our example, we will consider that an HRU area cannot exceed 2 ha.

Figure 5.4 shows an example of badly-shaped elements that must be improved through the segmentation procedure. This subcatchment forms part of the Mercier catchment that is the main case study of this chapter.

In the following, we describe the tool that allows the automatic segmentation of each badly-shaped element from GIS layers in the .shape format.

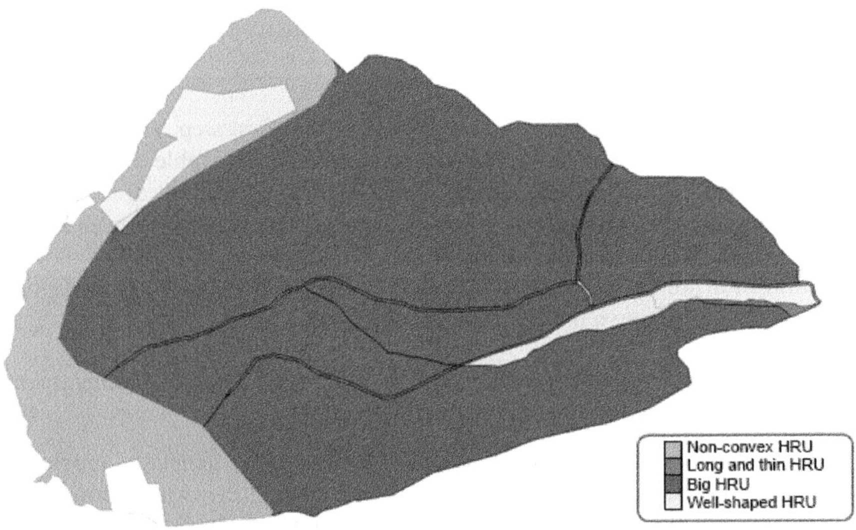

Figure 5.4. *Badly-shaped elements in the Mercier subcatchment. For a color version of the figure, see www.iste.co.uk/baghdadi/qgis4.zip*

5.2. Implementation of the TriangleQGIS module and general methodology

5.2.1. *Used technologies*

The use of GIS for preprocessing input information for a hydrological model has become a requirement, given the large amount of geographical information that must be manipulated. GIS platforms facilitate those treatments. The most used (among free platforms) are QGIS and GRASS. QGIS has a Graphical User Interface that can be incorporated into other GIS systems or databases and includes plugins developed by the users. All these reasons led to the decision to implement the Triangle plugin in QGIS. The plugin allows the use of the Triangle® software and working directly with vectorial layers in the .shape format, without needing to export layers to the native format of Triangle® (.node, .ele, .ploy, .area, .edge and .neigh files).

The Python language is widely used to implement plugins in QGIS and we used it as the most appropriate for the implementation of our plugin. In addition, thanks to Python libraries, it is possible to emulate the use of QGIS and GRASS-GIS and simplify the preprocessing step of distributed hydrological models.

5.2.2. *Context and general methodology*

The main motivation for developing the QGIS plugin presented in this section is the segmentation of highly non-convex polygons. In general, available triangulation algorithms allow the generation of triangles that form a convex closure, but that do not preserve boundaries of the original polygon. In addition, classical triangulations do not allow the imposition of restrictions on the area or angles of triangles that compose the polygon. These features are included in the Triangle® software, developed by Jonathan Shewchuck [SHE 96]. This software also provides the most efficient and robust triangulation method to improve our 2D model mesh. The TriangleQGIS plugin allows the activation of the Triangle® software from QGIS, using polygonal elements in the .shape format.

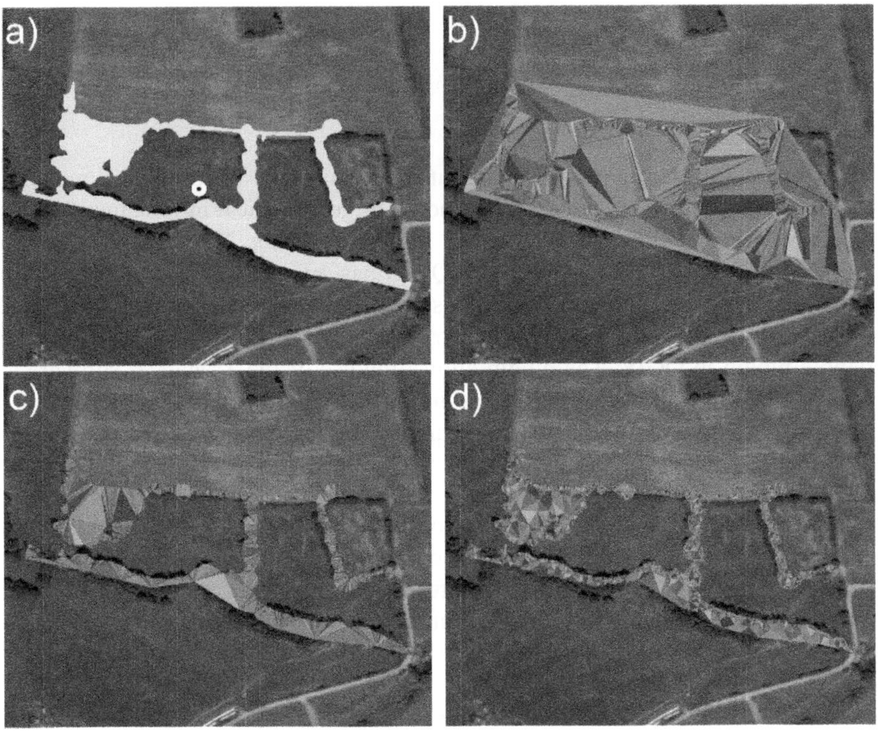

Figure 5.5. *a) HRU representative of a green area. b) Classical Delaunay triangulation: 4,667 triangles. c) Triangulation without restrictions on area using Triangle®: 2,340 triangles. d) Triangulation with angle restriction to 20°: 5,247 triangles. For a color version of the figure, see www.iste.co.uk/baghdadi/qgis4.zip*

QGIS tool to realize a triangulation:
- Triangulation tool: *Vector > Geometrical tools > Delaunay triangulation.*

Figure 5.5 shows an HRU associated with a green area and the results of its triangulation using various options of the software. Figure 5.5(b) shows the triangulation of the whole convex domain, known as "Delaunay" triangulation, obtained from the "vector" menu of QGIS. The drawback of this option is that the triangular mesh uses only nodes and can only generate a convex envelope. The original boundary of the polygon is lost. This is not the mesh we want to obtain, as we want to preserve the original polygon shape. Figure 5.5(c) shows the triangulation obtained using the Triangle® tool. The decomposition in triangles uses the original polygon, without generating the convex envelope. Figure 5.5(d) shows a triangulation also obtained with the Triangle® tool, but with an additional restriction of the maximum area of the triangles (A_{max} = 2 ha). To fulfill this criterion, additional nodes are introduced inside the polygon. This is automatically done using the Triangle® software to respect the imposed criteria on areas (or angles). The Triangle® software also proposes other options that allow putting restrictions on angles of the generated triangles for each polygon. For all these reasons, this software was very well suited to improve the geometry of the HRUs and was implemented in QGIS.

To decompose the polygons of our model mesh, we chose to generate a Delaunay triangulation with restrictions (using Triangle®). However, even if the elements obtained after this step are convex, there is a large number of triangles. This can be detrimental to the computing performance of the numerical model that will use this mesh. In addition, we would like to get polygons as close as possible to the shape of the initial polygons.

As a consequence, the triangulation step is followed by a step called "dissolution". It allows the preservation of the shape of HRUs, as elements of the landscape. The dissolution step (which is a grouping process) takes a triangle and combines it with its neighbor having the largest area. This process is continued until the maximum threshold for the criteria is exceeded. This dissolution step is applied to all triangles that have not been dissolved. Finally, this decomposition leads to final polygons that are more satisfactory for use in the hydrological model.

Polygons, the areas of which are larger than the maximum areas fixed by the user, may also be segmented in the same way as non-convex polygons that have their gravity center outside the polygon, using a restriction on the maximum area of the triangles, both in the initial triangulation phase and in the dissolution phase of those triangles.

In section 5.2.3, we present the work realized for the implementation and the execution of the plugin in QGIS (version 2.14 or below), complemented with the presentation of the use of some scripts from Geo-PUMMA in order to provide a whole view of the proposed methodology.

5.2.3. *Structure of the QGIS plugin*

To implement a plugin in QGIS, it is necessary to create some files and classes that implement some methods, or other files, containing all the relevant information for the plugin. A QGIS plugin called "Plugin Builder" already exists. It allows the user to generate the basic structure of a functional plugin. The files generated by this plugin are the following:

– *init.py*: this is the starting point of the plugin. Its role is only to initialize and set as output the main plugin;

– *metadata.txt*: this file contains the plugin metadata including: its name, the QGIS version required for its use, the author(s);

– *mainPlugin.py*: this class is the main class of the plugin, in which the specific functionalities of the plugin are implemented;

– *resources.qrc*: this file contains references to files, external to the plugin, like images or other files. It must be compiled using *pyrcc4* to produce the *resources.py* file, allowing the use of the external files by the plugin;

– *mainPlugin.ui*: this file contains the graphical interface of the main plugin. It must be compiled using *pyuic4* to get the *mainPlugin.py* file. This will allow the use of the plugin graphical interface;

– *mainPluginDialog.py*: the main objective of this class is the initialization of the plugin graphical interface.

5.2.4. *Basic used library: MeshPy*

One of the interesting characteristics of the MeshPy library is its ability to impose restrictions on the maximum area and on the minimum angle of the triangles that ensures the mesh quality.

This library is essential to the plugin as the Triangle® is incorporated into this library, so that it can interact with QGIS vectorial layers. Thus it allows the performance of triangulations with restrictions on the maximum surface or the minimum angle. Moreover, the library allows the generation of 2D meshes (with Triangle®) as well as 3D meshes (using tetrahedrons in TetGen12). The use of this library for the 2D triangulation is very similar to what would be done using Triangle® directly.

We also used the "Plugin Builder" plugin to get the basic structure of a functional plugin. We also mention the "Plugin Reloader" plugin that allows the immediate reloading of the plugin during its development, thus avoiding having to load QGIS each time the plugin must be loaded. To realize the graphical interface, we used QTCreator, which allows easy editing of the plugin graphical interface, called *mainPlugin.ui*. QTCreator is available here: https://www.qgis.org/en/site/getinvolved/development/qgisdevelopersguide/qtcreator.html .

5.2.5. *Installation of the plugin in Windows*

To run the TriangleQGIS plugin, the user must install QGIS (version 2.14 or below), Python 2.7 and the MeshPy library. The following steps were conducted on Windows XP and Windows 7.0 computers.

Install the following archives:

1) Quantum GIS: QGIS-OSGeo4W-2.4.0-1-Setup-x86_64;

2) Python: python-2.7.amd64;

3) MeshPy library: MeshPy-2014.1.win-amd64-py2.7.

Once these tools are installed, the folder "meshpy", located in C:Python27\Lib\site-packages, is copied into the folder C:Programes\QGIS Chugiak\apps\Python27\Lib\site-packages as shown in Figure 5.6.

It is necessary to verify in the Python windows of QGIS that the MeshPy library is running correctly. For that, the user can type the following command: "import meshpy". If there is no output message, it means that everything is running correctly. If not, previous steps must be verified. Figure 5.7 shows the Python windows and the result of the "import meshpy" command.

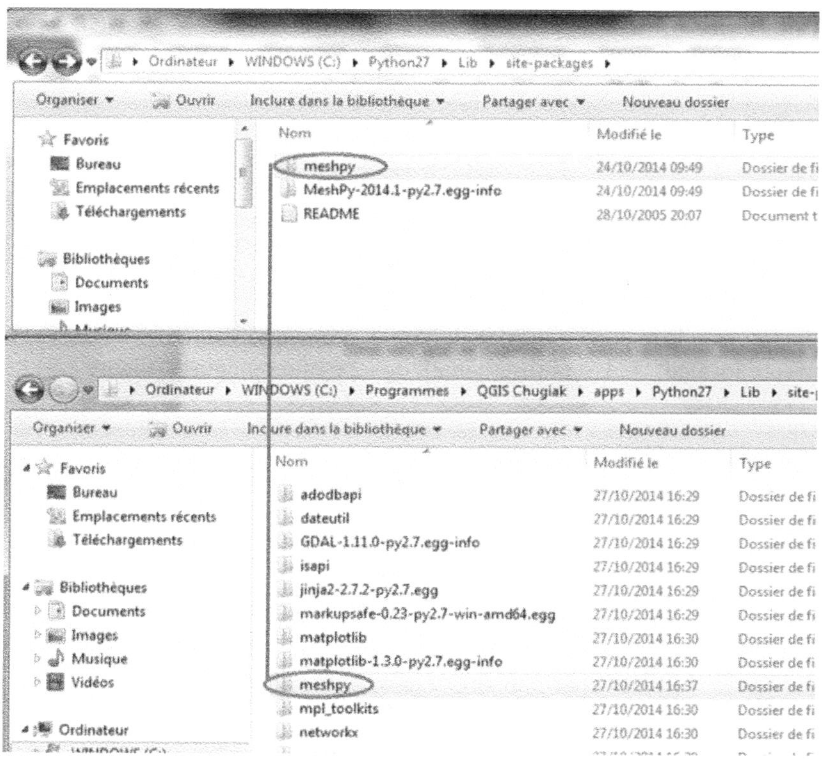

Figure 5.6. *Adding the MeshPy library to Python2.7. For a color version of the figure, see www.iste.co.uk/baghdadi/qgis4.zip*

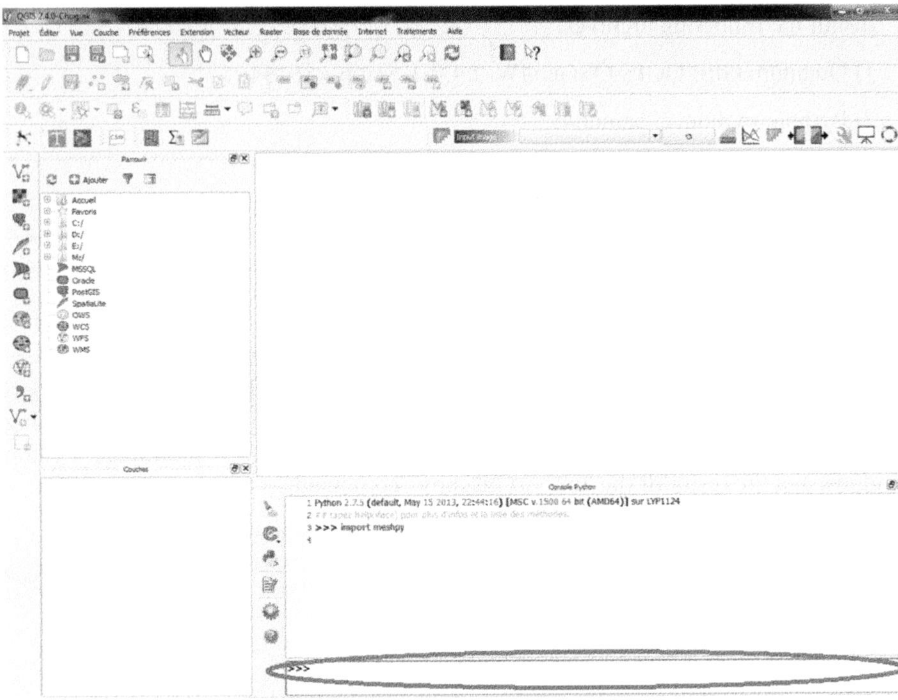

Figure 5.7. *Verification of the correct installation of the MeshPy library in the Python windows in QGIS*

The last step is to copy the TrianglePlugin archive to the folder where the QGIS plugins are located: "C:Users\[username]\.qgis2\python\plugins" (Figure 5.8).

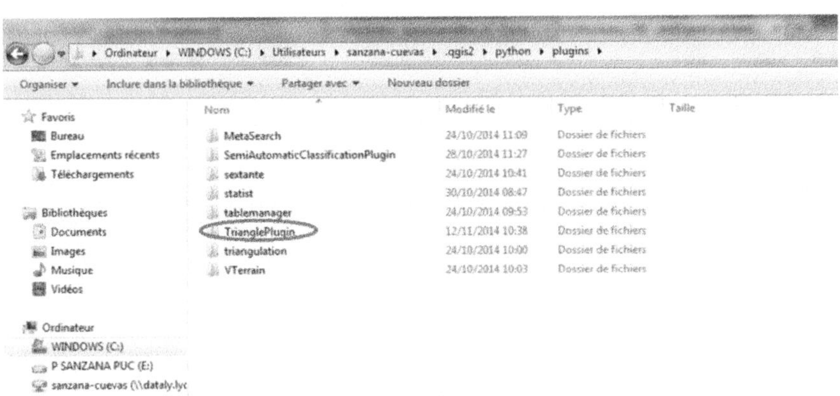

Figure 5.8. *Folder where the TrianglePlugin archive must be copied*

Finally, like any QGIS plugin, it must be activated from the QGIS toolbar by selecting Plugins>Manage and Install Plugins and activating TrianglePlugin (Figure 5.9).

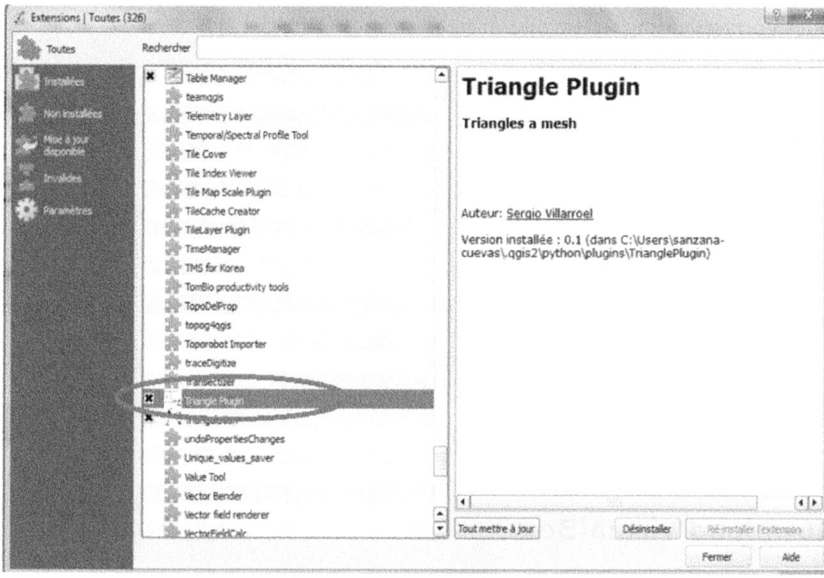

Figure 5.9. *Activation of the TriangleQGIS plugin*

Figure 5.10 shows the TriangleQGIS plugin interface.

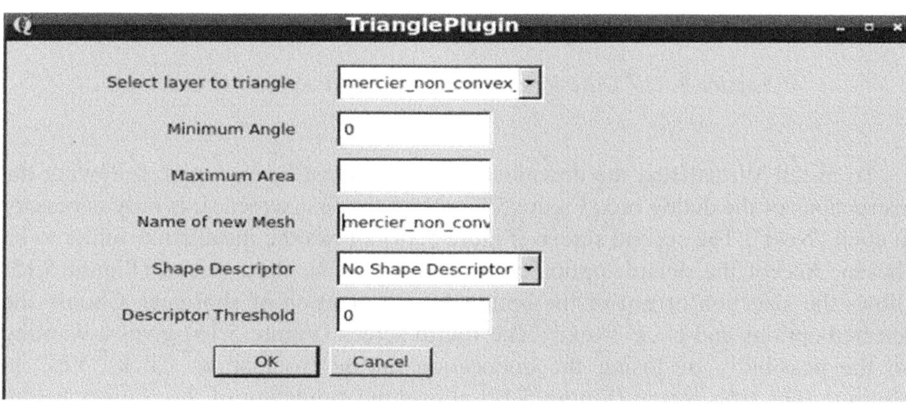

Figure 5.10. *The TrianglePlugin interface*

5.2.6. *Installation of the virtual box, QGIS plugin and Geo-PUMMA*

To install the “Virtual Box Machine”, it is recommended to visit the website (Figure 5.11) (https://www.virtualbox.org/) and to download, from the “download” tab, files corresponding to the operating system (Figure 5.12).

Figure 5.11. *Landing page of VirtualBox*

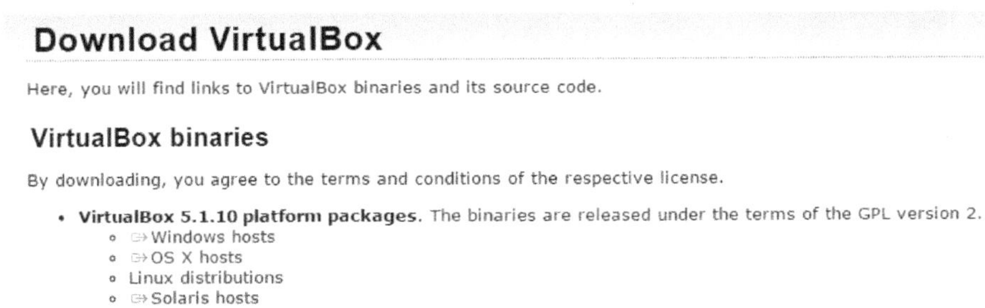

Figure 5.12. *“Download” section of the VirtualBox website*

To install VirtualBox, the downloaded archive must be executed, following the instructions of the dialog box. Figure 5.13 shows the first screen. It is only necessary to click “Next”. The second screen (Figure 5.14) allows the installation folder to be chosen. Accept the default options. Click “Next”. The third screen (Figure 5.15) allows the selection or not of the option for the creation of shortcuts. Choose the selected options and click “Next”. The fourth screen (Figure 5.16) gives a warning on the possibility of losing the connection during installation. Click “Yes” to continue. The fifth screen (Figure 5.17) allows the validation of the choices before launching the installation. If everything is OK, click “Install”. Windows users may receive a warning similar to the one shown in Figure 5.18. Click on “Install” to continue the installation process.

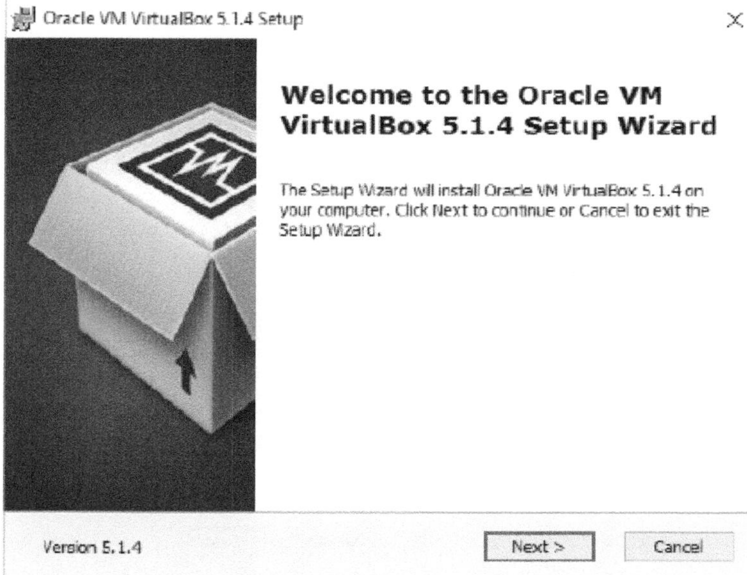

Figure 5.13. *First screen of the installation dialog box*

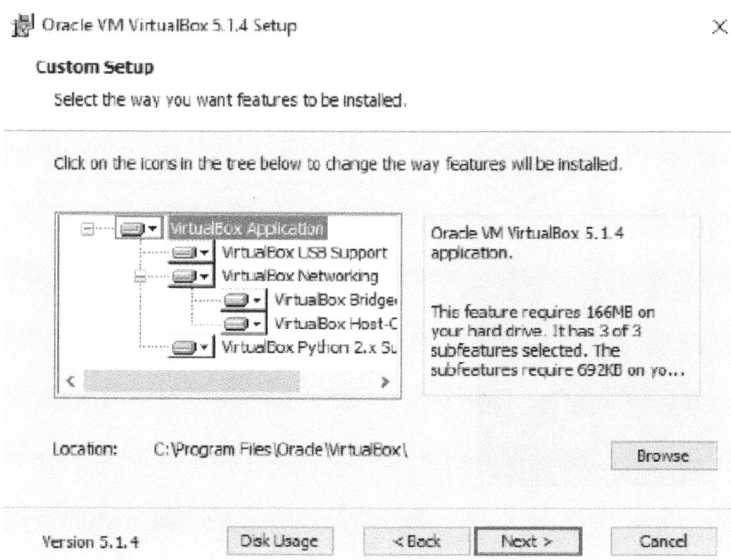

Figure 5.14. *Second screen of the installation dialog box*

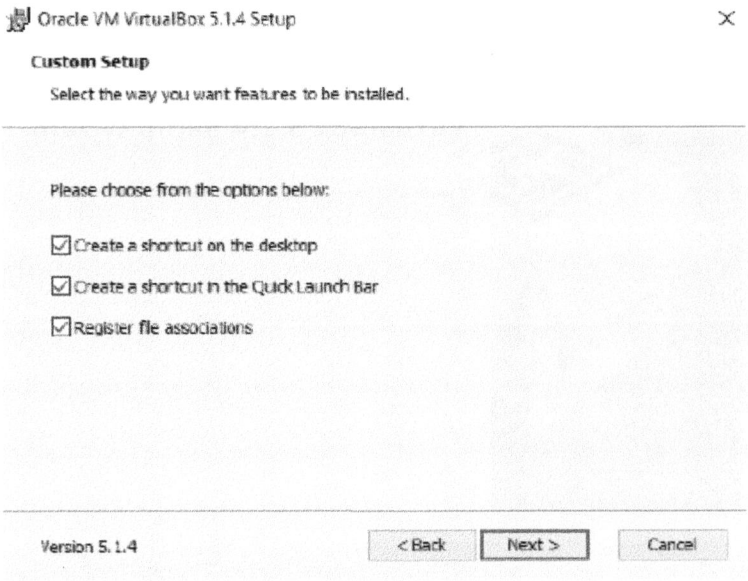

Figure 5.15. *Third screen of the installation dialog box*

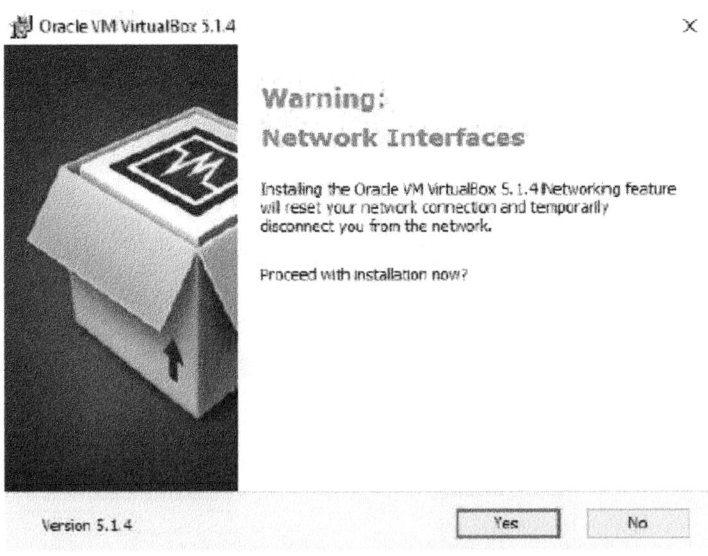

Figure 5.16. *Fourth screen of the installation dialog box*

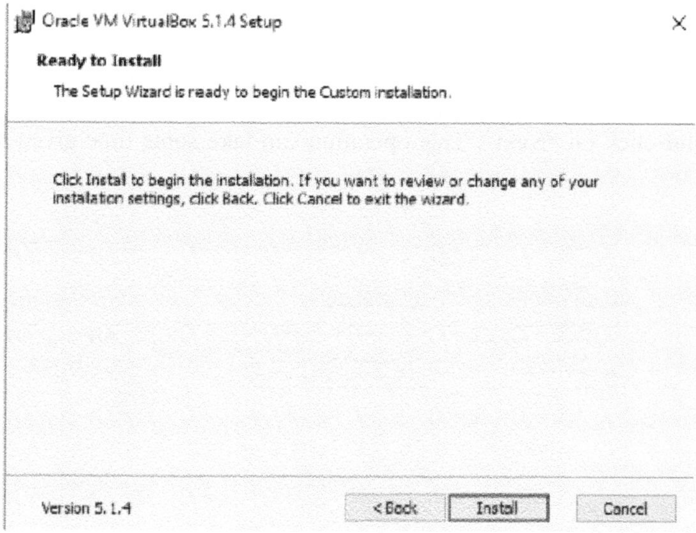

Figure 5.17. *Fifth screen of the installation dialog box*

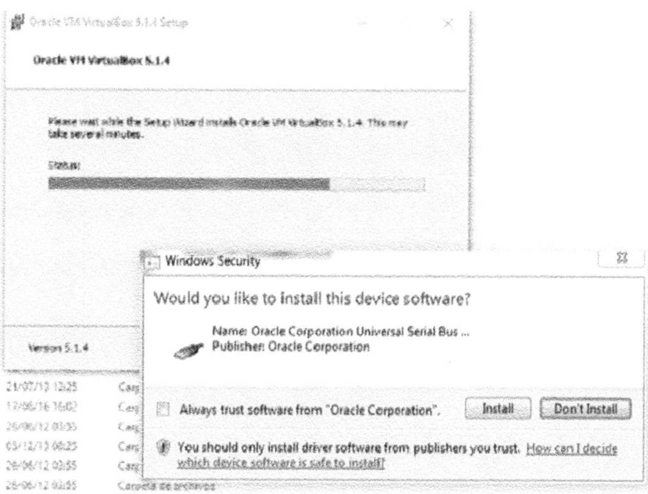

Figure 5.18. *Sixth screen of the installation dialog box*

Once the installation is terminated and successful, open VirtualBox (Figure 5.19).

What follows is related to the virtual box installation, allowing the running of Geo-PUMMA. The source of the scripts are available here[2] and the virtual box installation package here[3]. For that, click on "File", then select "Import Virtual Service". A screen similar to that shown in Figure 5.20 appears. Select the Virtual Box file and click on "Next". This operation can take some time given the size of the file.

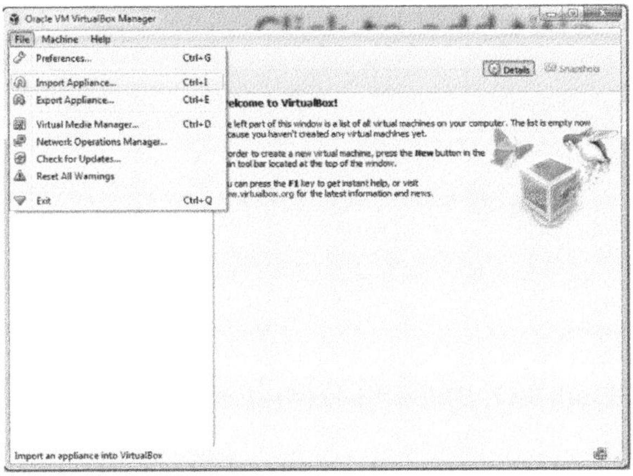

Figure 5.19. *Main screen of Virtual Box, option File ==> Import Appliance*

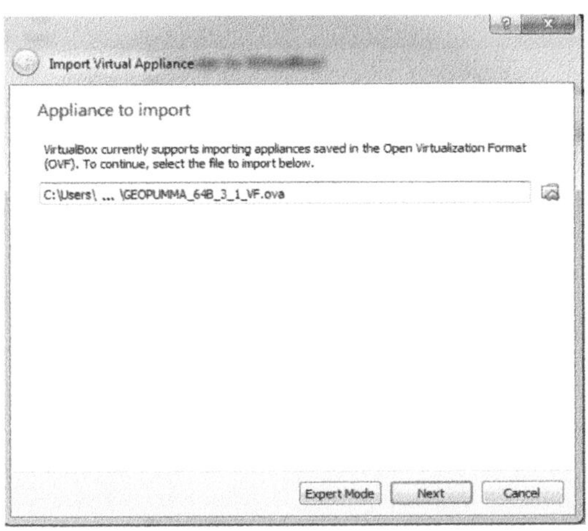

Figure 5.20. *Screen to import virtual services*

2 https://forge.irstea.fr/projects/geopumma.

3 http://doi.org/10.5281/zenodo.821563.

Once the virtual machine is loaded, it is necessary to verify if the operating system allows the use of 64-bit virtual boxes. If this is not the case, only the Ubuntu option (32-bit) will appear (see Figure 5.21 for the details of the verification of the existence of a problem or not).

If the 64-bit option does not appear, try the following commands:

1) switch off the computer;

2) switch it on again and immediately press the ESC or F10 button to access the BIOS (some computers allow the access to this functionality with another button);

3) in the BIOS menu, activate the option that allows the use of virtual machines;

4) accept and restart after saving the modifications.

Once the computer is switched on again, the Ubuntu option (64-bit) should be available.

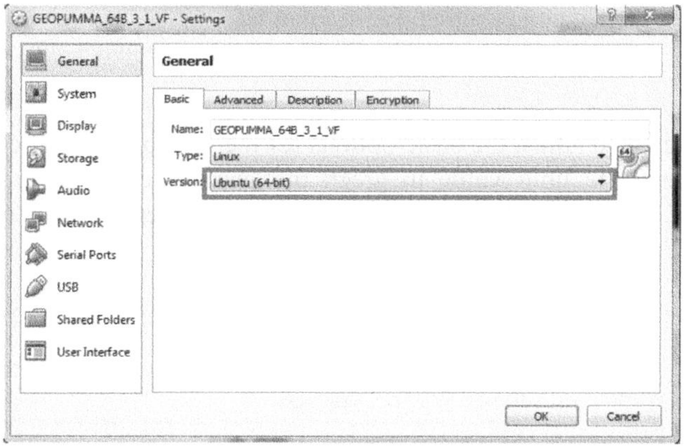

Figure 5.21. *Window for the configuration of the virtual box*

The disk space allocated to the virtual box cannot be accessed from the original operating system. However, it is possible to configure a shared folder that will be accessible from the operating system and from the virtual box. To do that, in the configuration menu, go to "Shared folders" and select the desired location by clicking on the chosen folder. Create a folder named "name_folder" on the hard disk and select it. Select also the option "Auto-mount" and "make permanent". Then click "OK" to close the window and click again on "OK" to add the folder (see Figure 5.22 for the visualization of these steps).

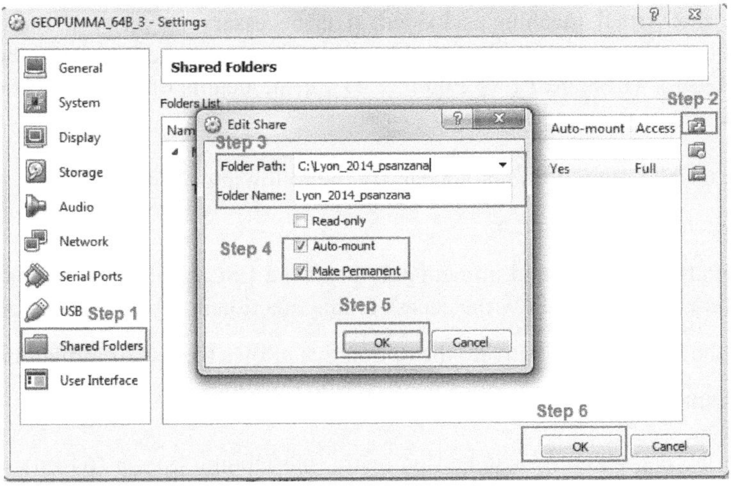

Figure 5.22. *Add a folder to the hard disk, shared between the operating system and the virtual box. For a color version of the figure, see www.iste.co.uk/baghdadi/qgis4.zip*

The last step is to adjust the basic options for the memory. The allocated memory must not exceed the limit value in green that varies according to the computer characteristics (Figure 5.23). It is also necessary to activate the option for 3D acceleration (Figure 5.24).

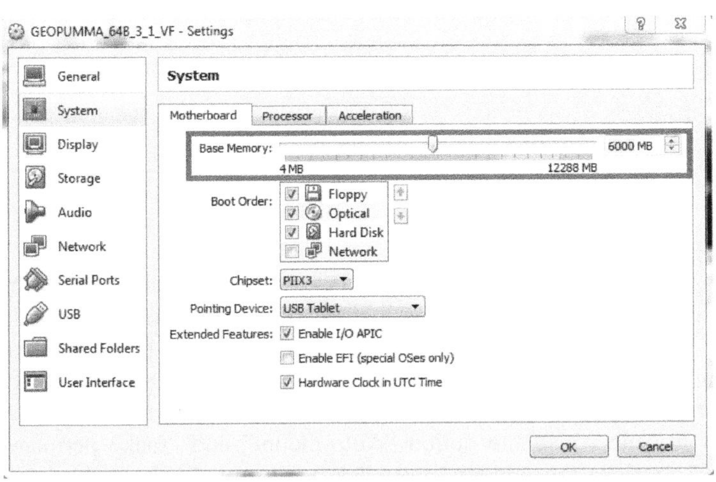

Figure 5.23. *Screen for memory allocation. For a color version of the figure, see www.iste.co.uk/baghdadi/qgis4.zip*

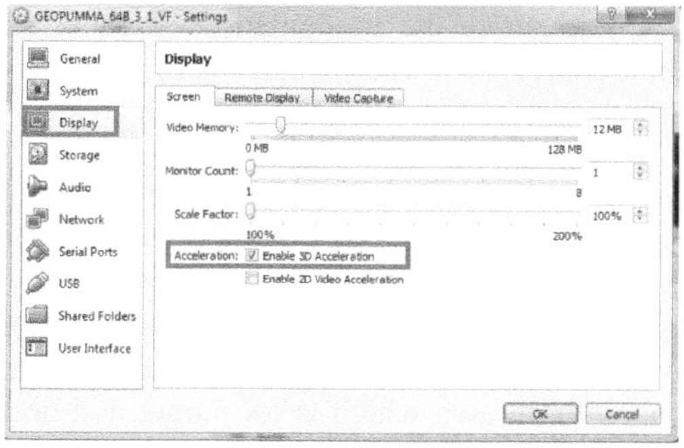

Figure 5.24. *Option to activate 3D acceleration. For a color version of the figure, see www.iste.co.uk/baghdadi/qgis4.zip*

Then, click to show the launch screen of the virtual box (Figure 5.25).

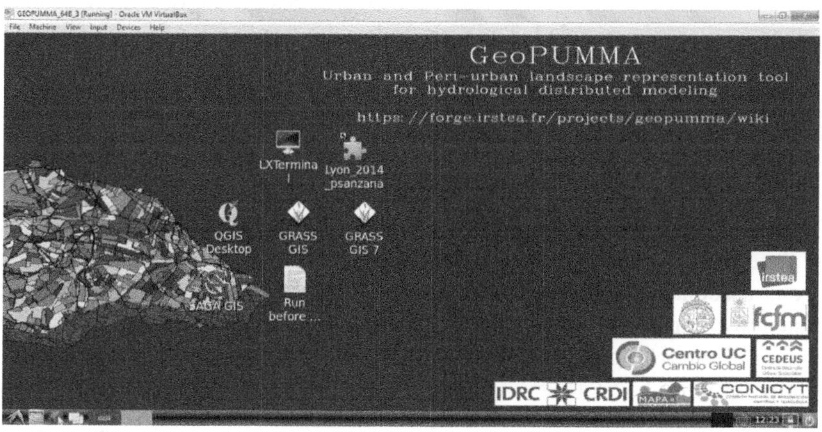

Figure 5.25. *Welcome screen of the Geo-PUMMA virtual box*

5.3. Illustration of the TriangleQGIS plugin and some Geo-PUMMA scripts

Explanations provided in this chapter allow an understanding of how the tools are working. If the user wishes for more details on the use of a specific GRASS or QGIS command presented in this chapter, they can refer to the Geo-PUMMA User's manual available at this link[4] that provides details on the various steps.

---

4 https://forge.irstea.fr/projects/geopumma/files.

The following steps illustrate the triangulation of badly shaped polygons. As explained in the methodology, it is first necessary to select non-convex elements, elements which are too long and thin, and elements with an area which is too large.

The triangulation step is important to improve the hydrological model mesh. That is why, according to the chosen improvement criteria (non-convexity, maximum area, or long and thin elements), it could be necessary to add nodes to get a satisfactory triangulation. This is not necessary when polygons are triangulated because of the convexity criterion. On the other hand, nodes are added inside the polygon for polygons with an area which is too large, and nodes are added at the polygon edges in order to get smaller triangles for long and thin polygons.

5.3.1. *Insertion of nodes for long and thin polygons*

The method to add nodes at the edges of long and thin elements is described here. A maximum distance of 5 m between two nodes on the boundary of long and thin elements is recommended. To insert nodes, the *v.split* GRASS command can be used as explained below.

```
v.split input=mercier_long_elements output=mercier_long_elements_split length=5
```

This way, we ensure that the triangulation will lead to small-sized triangles that allow us to build satisfactory polygons in a second step. Figure 5.26 shows the initial polygon, associated with a road segment (Figure 5.26(a)) with nodes initially separated with irregular distances (Figure 5.26(b)). After inserting nodes with a maximum distance, a polygon with nodes separated by a maximum distance of 5 m is obtained (Figure 5.26).

Figure 5.26. *a) Polygon associated with a road segment, b) with the initial nodes and c) the newly inserted nodes in red that ensure a maximum distance of 5 m separating the nodes. For a color version of the figure, see www.iste.co.uk/baghdadi/qgis4.zip*

GRASS function to insert nodes:
- v.split

QGIS tool to verify the nodes:
- Triangulation tools: *Vector > Geometric tools > Nodes extraction*

5.3.2. *Triangulation using the TriangleQGIS plugin*

The TriangleQGIS plugin can be used directly with vectorial layers in the .shape format. These layers can be obtained from various GISs (ArcGIS, GRASS-GIS, OpenJump, OrbiGIS, etc.). We recommend saving then again in the QGIS .shape format, using options described in Figures 5.27. It is recommended to do this step for all the files that contain elements selected from the convexity criteria (*mercier_non_convex*), the maximum area criterion (*mercier_big_areas*) and the long and thin elements criterion (*mercier_long_elements*). Figure 5.27 shows the steps to follow (1–5 to save the files in the vectorial format).

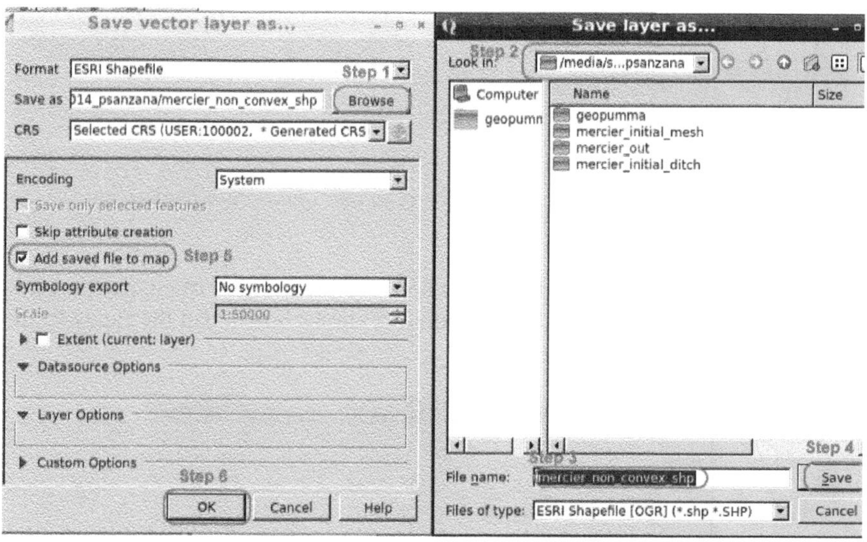

Figure 5.27. *Steps to save a file in the QGIS vectorial format. For a color version of the figure, see www.iste.co.uk/baghdadi/qgis4.zip*

First, the identifiers of the badly-shaped elements must be loaded. Elements selected from the convexity criterion are triangulated first. Figure 5.28 shows the selected badly shaped elements from the convexity criterion.

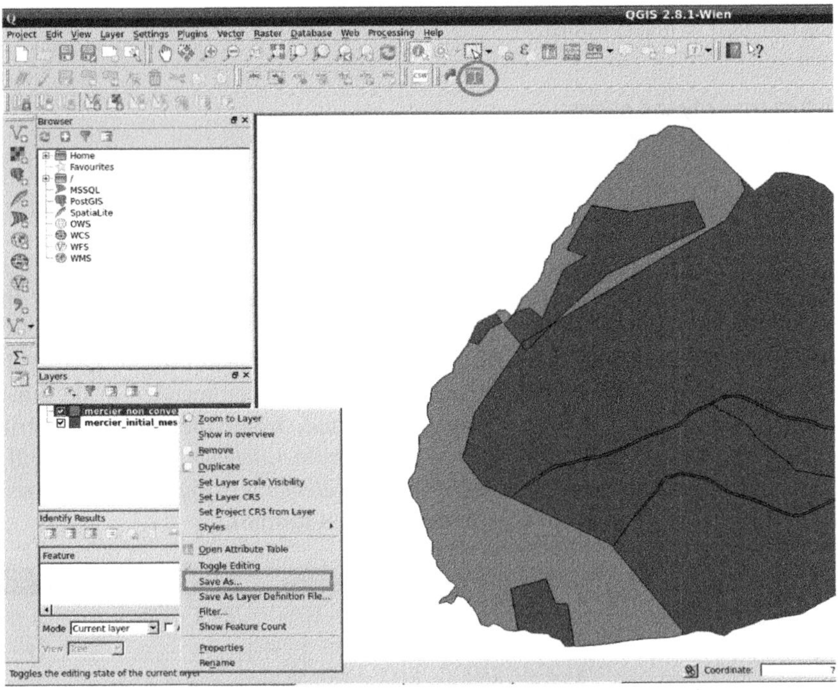

Figure 5.28. *HRU selected with a Convexity Index = 0.590. For a color version of the figure, see www.iste.co.uk/baghdadi/qgis4.zip*

Figure 5.29 shows the interface for choosing the various triangulation options available in the TriangleQGIS plugin. Table 5.2 shows how to fill the fields for non-convex elements.

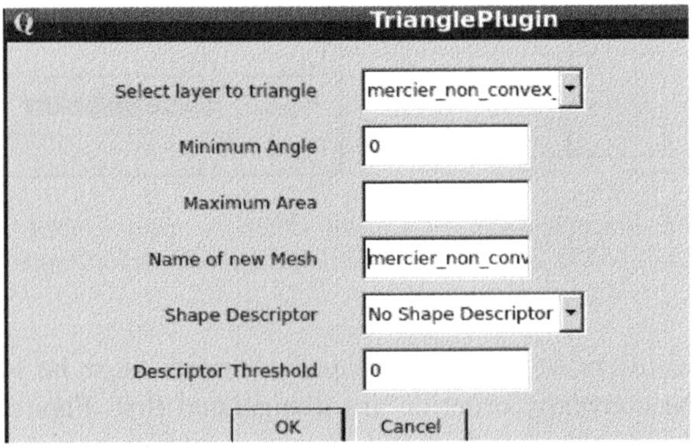

Figure 5.29. *Interface of the TriangleQGIS plugin*

Case	Non-convex elements layer. No constraint in the triangulation	Non-convex elements layer. Triangulation with constraint on the maximum area 20,000 m^2	Non-convex elements layer. Triangulation with constraint of 30° on the minimum angle
Input GIS layer to triangle	mercier_non_convex_shp	mercier_non_convex_shp	mercier_non_convex_shp
Minimum angle	0	*Keep empty*	30
Maximum area	*Keep empty*	20,000	*Keep empty*
Name of the new mesh	mercier_non_convex_t	mercier_non_convex_area	mercier_non_convex_angle
Shape descripted	No shape descriptor	No shape descriptor	No shape descriptor
Descriptor threshold	0	0	0

Table 5.2. *Options to generate the various triangulations with the TriangleQGIS plugin (to fill the required fields in the graphical interface)*

Table 5.3 shows the result of each triangulation. For the new generated vectorial mesh, comparison criteria are the number of triangles and statistics on areas and minimum angle. In the first example, a triangulation without restriction on the maximum area and minimum angle was selected. In this case, the triangulation uses only existing nodes on the initial boundary (the total number of obtained triangles is 174). In the second example, the triangulation is realized using a restriction on the maximum area of 2 ha. In this case, the program inserts new nodes inside the polygon so that the generated triangle size is lower than 2 ha. The total number of triangles is 203 triangles. In the third example, the triangulation is realized with a restriction on the angles that must be larger than 30°. This requires inserting new nodes inside the polygon as previously. The total number of triangles is 814 triangles. The number of triangles generated from a single polygon can dramatically increase the final number of elements in the model. We see that up to 800 triangles can be generated.

Although the triangulation with restriction on the area (Figure 5.30(c)) or restriction on the angles (Figure 5.30(d)) leads to well-shaped elements, the final element number is very large and includes very small triangles, so that the final result cannot be considered as satisfactory for the modeling. The method providing the best compromise between the number of final elements and "good shape" was finally retained, i.e. the triangulation without restriction (Figure 5.30(b)).

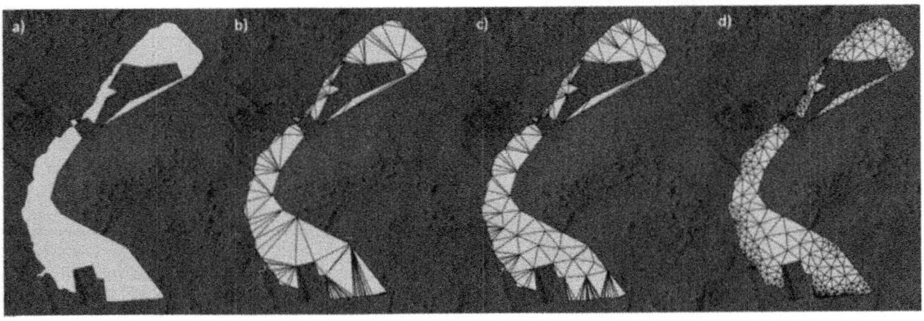

Figure 5.30. *a) Non-convex polygon, b) triangulated without restriction, c) triangulated with restriction on the maximum area (2 ha) and d) triangulated with restriction on the minimum angle 30°*

Polygon	No. Δ	Min area m^2	Max area m^2	Median area m^2	Min angle °
Initial (Figure 5.30(a))	-	-	-	89.35	11
Triangulated without restriction (Figure 5.30(b))	174	0.17	6.618	510	1.6
Triangulated with restriction on the maximum area (Figure 5.30(c))	203	1.58	1.928	437	2.1
Triangulated with restriction on the minimum angle (Figure 5.30(d))	814	0.12	1.417	109	30

Table 5.3. *Number of triangles (Δ), area and angles for the element in Figure 5.30*

For long and thin elements, generally associated with road segments, the triangulation step is also relevant. Possible options for the triangulation are described in Table 5.4. Figure 5.31 shows an HRU associated with a road segment. As the Mercier subcatchment is rather rural, we use here a .shp example taken in the El Guido subcatchment in Chili. Initially, with the options of Table 5.4, this road segment is represented by an ensemble of nodes (Figure 5.31(a)). Then a triangulation according to various criteria is realized: a triangulation without restriction (Figure 5.31(b)), with restriction on the area that must not exceed 200 m^2 (Figure 5.31(c)) and with restriction on the angles that must be larger than 30° (Figure 5.31(d)). In a second step, nodes with a maximum distance of 5 m are inserted in the polygon initial boundary and a triangulation without restriction is performed.

Case	Triangulation without restriction (initial nodes)	Long and thin element triangulation with restriction on the maximum area of 200m^2 (original nodes)	Long and thin element triangulation with restriction on the minimum angle of de 30° (original nodes)	Long and thin element triangulation without restriction (nodes with d_{max}= 5m)
Input GIS layer to Triangle	test_slim_long_split_dmax_5m	test_slim_long_split_dmax_5m	test_slim_long_split_dmax_5m	test_slim_long_split_dmax_5m
Minimum angle	0	*0*	30	0
Maximum area	*Keep empty*	*200*	*Keep empty*	*Keep empty*
Name of the new mesh	test_slim_long_T	test_slim_long_T	test_slim_long_T	test_slim_long_T
Shape descriptor	No shape descriptor	No shape descriptor	No shape descriptor	No shape descriptor
Descriptor threshold	0	0	0	0

Table 5.4. *Options for the triangulation of long and thin elements using the TriangleQGIS plugin (to fill out information required in the graphical interface fields)*

Figure 5.31. *a) HRU associated with a road segment with the original nodes. b) Triangulation without restriction. c) Triangulation with restriction on the maximum area of 200 m². d) Triangulation with restriction on the minimum angle of 30°. e) HRU where nodes are separated with a maximum distance of 5 m (f) and its triangulation without restriction on the area or on the angles. For a color version of the figure, see www.iste.co.uk/baghdadi/qgis4.zip*

Table 5.5 shows the results obtained for each triangulation.

Polygon	No. Δ	Min area m^2	Max area m^2	Median area m^2	Min angle °
Initial (Figure 5.31(a))	-	-	-	3,439	71
Triangulated without restriction (Figure 5.31(b))	105	0.07	467	32.7	4.4
Triangulated with restriction on the maximum area (Figure 5.31(c))	218	0.01	201	15.7	21.1
Triangulated with restriction on the minimum angle (Figure 5.31(d))	476	0.01	78	7.2	30
Triangulated with restriction and after insertion of nodes (Figure 5.31(f))	278	0.01	125	12.3	1.2

Table 5.5. *Number of triangles (Δ), area and angles for the element in Figure 5.31*

The triangulation without restriction of the polygon leads to very long polygons, because the distance between nodes is not controlled by the user (Figure 5.31). Triangulation with restriction on the maximum area (Figure 5.31(c)) and on the minimum angles (Figure 5.31(d)) introduces new nodes inside the polygon that will not lead to good results once these triangles are dissolved according to the Form Factor (see section 5.3.3). The polygon that is triangulated after insertion of nodes on the boundary leads to smaller and equidistant edges, better suited for the subsequent dissolution step according to the Form Factor criterion (section 5.3.3).

For very large elements, the segmentation is realized with a constraint on the maximum area. Figure 5.32 shows an HRU representative of a polygon with large area, 19 ha, that is triangulated according to different options: without any restriction (Figure 5.32(b)), restriction on the maximum area fixed to 2 ha (Figure 5.32(c)) and restriction on the minimum angle fixed to 30° (Figure 5.32(d)). Table 5.6 shows the options that must be chosen to run these various triangulations and Table 5.7 presents the results with statistics associated with each triangulation.

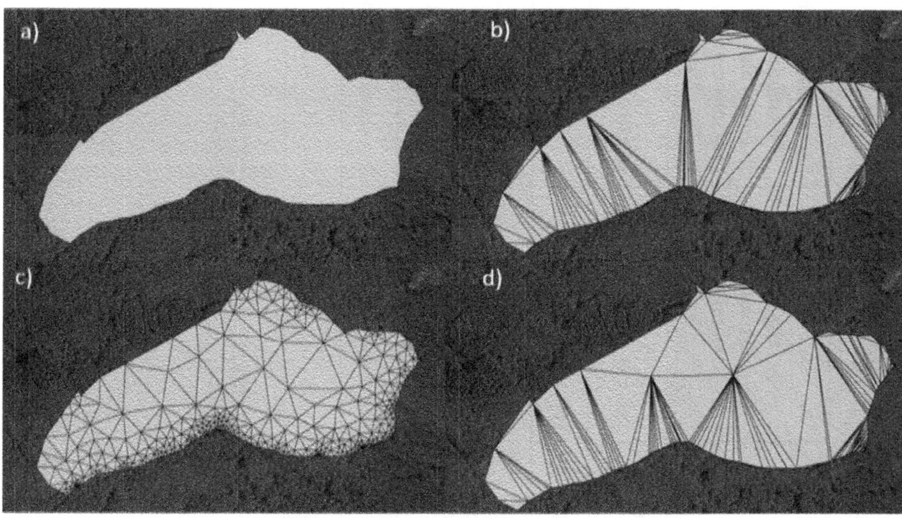

Figure 5.32. *a) Very large polygon, b) triangulated without restriction, c) triangulated with restriction on the maximum area (2 ha), and d) triangulated with restriction on the minimum angle 30°*

Case	Very large element. No constraint in the triangulation	Very large element. Triangulation with constraint on the maximum area 20,000 m^2	Very large element. Triangulation with constraint of 30° on the minimum angle
Input GIS Layer to Triangle	mercier_big_area_shp	mercier_big_area_shp	mercier_big_area_shp
Minimum angle	0	*0*	30
Maximum area	Keep empty	20,000	Keep empty
Name of the new mesh	mercier_big_area_t	mercier_big_area_area	mercier_big_area_angulo
Shape descriptor	No shape descriptor	No shape descriptor	No shape descriptor
Descriptor threshold	0	0	0

Table 5.6. *Options to generate the triangulation of a very large polygon with the TrangleQGIS plugin (to fill out the required fields in the graphical interface)*

	No. Δ	Min area m^2	Max area m^2	Median area m^2	Min angle °
Initial (Figure 5.32(a))	-	-	-	192.14	22.8
Triangulated without restriction (Figure 5.32(b))	160	0.45	24.27	1.19	1
Triangulated with restriction on the maximum area (Figure 5.32(c))	164	0.45	19.34	1.16	1
Triangulated with restriction on the minimum angle (Figure 5.32(d))	841	0.09	5,532	228	30

Table 5.7. *Number of triangles (Δ), area and angles for the very large element in Figure 5.32*

The number of triangles obtained after the triangulation without restriction is 160, with a maximum area of 24,271 m^2. This decomposition does not fulfill the maximum area restriction of 20,000 m^2 and it is necessary to insert nodes inside the polygon to generate decomposition with smaller triangles. The triangulation with restriction on the maximum area results in 164 triangles that indeed lead to a maximum area of 2 ha. The triangulation with restriction on the angles leads to more regular polygons but their number increases up to 841 triangles. The most appropriate segmentation is therefore the second one, with restriction on the maximum area.

In summary, from the presented examples, the following recommendations can be made for the various types of badly-shaped elements:

1) non-convex HRUs: triangulation without restriction on areas and angles;

2) very long and thin HRUs: triangulation without restriction on areas and angles, but starting from a polygon where the distance between boundary nodes is lower than 5 m;

3) very large HRUs: triangulation with restriction on the maximum area with a typical value of 2 ha (new nodes are inserted inside the polygon to fulfill this criterion).

GRASS tools to insert nodes:
- v.split

QGIS tools to verify nodes:
- Geometrical treatment tools: *Vector > Geometry tools > Node extraction*

QGIS tool to perform triangulation:
- TriangleQGIS plugin

Figure 5.33 shows the triangulation of all the polygons that required it in the Mercier case study subcatchment (see Figure 5.4).

The use of the plugin implemented in QGIS allows us to use the Triangle® software directly on vectorial layers. It can be used to perform a Delaunay triangulation or triangulation with constraints, as shown in the previous figures. Our case study deals with the improvement of a distributed hydrological model mesh, but this plugin can also be used for other types of meshes requiring improvement of the elements' shape.

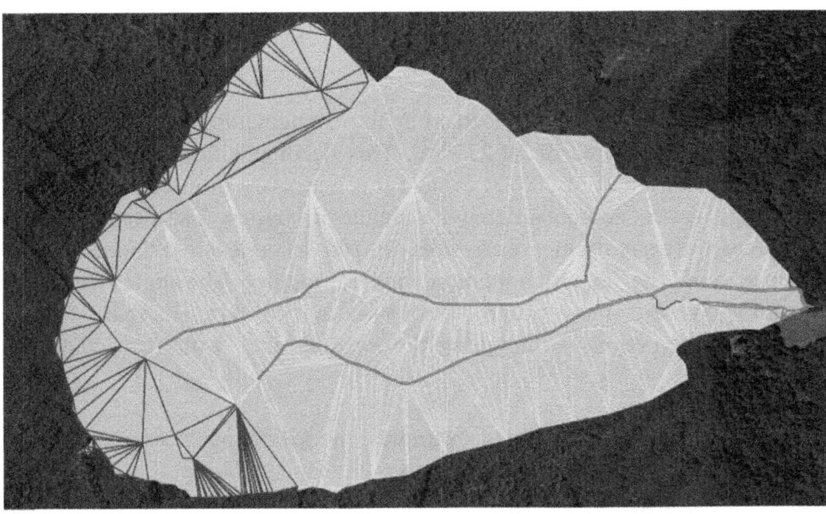

Figure 5.33. *Mesh where all the badly-shaped elements have been segmented using the TriangleQGIS plugin*

5.3.3. *Dissolution of triangulated elements*

Once the most appropriate triangulation tool has been selected to segment each element, dissolution of the obtained triangles can be performed using scripts from the Geo-PUMMA toolbox. This dissolution step allows us to rebuild pseudo-regular polygons that respect convexity and form factor criteria. The final result is therefore closer to the initial polygons, with a smaller number of elements than if all the triangles were kept. This step cannot be performed in QGIS and requires the use of the Geo-PUMMA toolbox and GRASS functions. These steps are however presented for completeness of the presented example. The interested reader can refer to Sanzana *et al.* [SAN 17] and to the Geo-PUMMA User's Manual [GEO 17] for further details.

The script allowing the segmentation of non-convex elements is:

```
p.B8.a.convexity_segmentation.py
```

Script:	p.B8.a.convexity_segmentation.py
Inputs:	• Mercier_non_convex_triangle layer (clean topology) • Mercier_non_convex layer (pre-Triangle) • A threshold convexity index • A maximum area value
Output:	• Mercier_non_convex_processed

Figure 5.34. shows the result of the segmentation using this script.

Figure 5.34. *Polygons dissolved using a convexity criterion (CI > 0.75)*

The script allowing the segmentation of elements according to the Form Factor criterion is:

```
p.B8.b.formfactor_segmentation
```

Script:	p.B8.b.formfactor_segmentation.py
Inputs:	• Mercier_long_elements_triangle layer • A maximum area value • A threshold value for the Form Factor
Output:	• Mercier_ long_elements _processed

Figure 5.35 shows the result of the segmentation using this script.

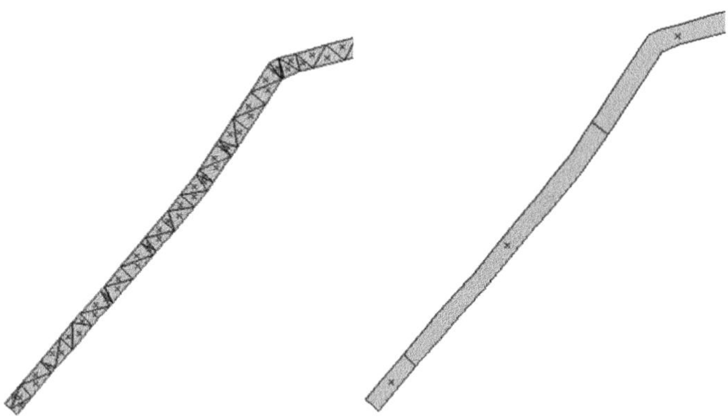

Figure 5.35. *Polygon dissolved according to the Form Factor criterion FF = 0.20*

The script allowing the segmentation of very large elements is:

```
p.B8.a.convexity_segmentation.py
```

Script:	p.B8.b.formfactor_segmentation.py
Inputs:	• Mercier_big_areas_triangle layer • Mercier_ big_areas layer (pre Triangle) • A threshold on the convexity index • A maximum area value
Output:	• Mercier_ big_areas _processed

Figure 5.36 shows the result of the segmentation using this script.

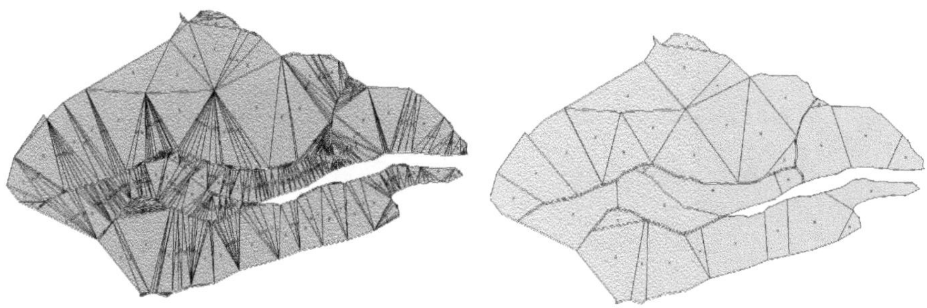

Figure 5.36. *Polygon dissolved according to the maximum area criterion Amax = 2 ha*

It can be seen that the dissolution dramatically reduces the final number of polygons required to represent HRUs. These elements, once segmented into pseudo-regular elements, allow a routing of the flow of water from each element to the catchment outlet. This routing tree will be satisfactory as illustrated in section 5.3.4.

5.3.4. *Effect of the model mesh improvement*

In a mesh generated to represent hydrolandscapes, the segmentation procedure will directly affect the hydrological connectivity. The flow routing from one element up to the catchment outlet can be obtained using the *p.olaf.py* script that also belongs to the Geo-PUMMA toolbox [SAN 17]. Flow routing is computed for each HRU and follows the topography toward the lowest neighbor or a river reach. Figure 5.37 shows one example for a subcatchment of the Mercier catchment. The figure shows the initial segmentation and topography contour lines (Figure 5.37(a)), the flow routing using the initial mesh (Figure 5.37(b)), the model mesh after improving non-convex elements only (Figure 5.37(c)) and the model mesh once all the elements requiring it have been improved (Figure 5.37(d)).

Topology problems are solved using the proposed segmentation, even if the final number of elements is increased as compared to the initial mesh. However, the segmentation of non-convex and very thin and long elements with threshold values of 0.75 for the Convexity Index and 0.2 for the Form Factor leads to a good compromise between accuracy and the increase in the number of elements.

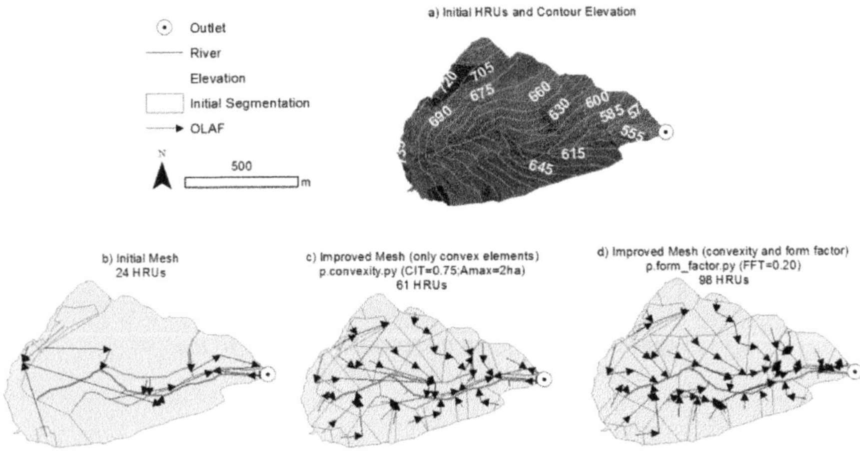

Figure 5.37. *a) Subcatchment in the Mercier and topography contour lines. Illustration of the effect of badly shaped elements improvement. b) Extraction of the drainage network using the initial mesh, c) the mesh where only non-convex elements have been improved, and d) mesh where all the badly-shaped elements are segmented. "OLAF" arrows correspond to flow direction between model elements. For a color version of the figure, see www.iste.co.uk/baghdadi/qgis4.zip*

5.4. Acknowledgments

The plugin was developed thanks to a program in support of PhD theses at the Vicerrectoría Académica de la Pontificia Universidad Católica. This VRI-UC project led to the production of a manuscript written in Spanish and entitled: "Introducción al Trabajo de Titulo CC6908: Generación automática de mallas 2D para modelar cuencas periurbanas mediante el concepto de Unidades de Respuesta Hidrológica (URHs) usando QGIS". This manuscript was written by Sergio Villarroel, student of the University of Chile, under the supervision of Nancy Hitschfeld, from the Computing Sciences Department. The authors were funded by the ECOS-CONICYT C14U02. We also thank the FONDECYT No. 1131131, CEDEUS (FONDAP 15110020), FONDECYT ENL009/15 and IRSTEA-Lyon for their support. Jorge Gironás acknowledges CIGIDEN (FONDAP 15110017). The Mercier catchment is part of OTHU (Observatoire de Terrain en Hydrologie Urbaine). This work was partially developed within the framework of the Panta Rhei Research Initiative of the International Association of Hydrological Sciences.

5.5. Bibliography

[BOS 12] BOCHER E., MARTIN J.Y., "TAnaTo2: A Tool to Evaluate the Impact of Natural and Anthropogenic Artefacts with a TIN-Based Model", in BOCHER E., NETELER M. (eds), *Geospatial Free and Open Source Software in The 21st Century*, Springer, Berlin, 2012.

[FLU 95] FLÜGEL W.A., "Delineating hydrological response units by geographical information system analyses for regional hydrological modelling using PRMS/MMS in the drainage basin of the River Bröl, Germany", *Hydrological Processes*, vol. 9, nos. 3–4, pp. 423–436, 1995.

[FUA 17] FUAMBA M., BRANGER F., BRAUD I. *et al.*, "Value of distributed water level and soil moisture data in the evaluation of a distributed hydrological model: Application to the PUMMA model in the semi-rural Mercier catchment (6.6 km^2) in France", *Journal of Hydrology*, 2017.

[GEO 17] GEO-PUMMA TEAM, Geo-PUMMA Tutorial v.1, available at: https://forge.irstea.fr/projects/geopumma, 2017.

[GIR 10] GIRONÁS J., NIEMANN J.D., ROESNER L.A. *et al.*, "Evaluation of methods for representing urban terrain in stormwater modeling", *Journal of Hydrologic Engineering*, vol. 15, pp. 1–14, 2010.

[JAN 11] JANKOWFSKY S., Understanding and modelling of hydrological processes in small peri-urban catchments using an object oriented and modular distributed approach. Application to the Chaudanne and Mercier sub-catchments (Yzeron Catchment, France), PhD Thesis, University of Grenoble, France, available at: http://tel.archives-ouvertes.fr/tel-00721988, 2011.

[JAN 13] JANKOWFSKY S., BRANGER F., BRAUD I. *et al.*, "Comparison of catchment and network delineation approaches in complex suburban environments: application to the Chaudanne catchment, France", *Hydrological Processes*, vol. 27, no. 25, pp. 3747–3761, 2013.

[JAN 14] JANKOWFSKY S., BRANGER F., BRAUD I. *et al.*, "Assessing anthropogenic influence on the hydrology of small peri-urban catchments: Development of the object-oriented PUMMA model by integrating urban and rural hydrological models", *Journal of Hydrology*, vol. 517, pp. 1056–1071, 2014.

[LAF 06] LAFONT M., VIVIER A., NOGUEIRA S. *et al.*, "Surface and hyporheic oligochaete assemblages in a French suburban stream", *Hydrobiologia*, vol. 564, pp. 183–193, 2006.

[LAG 10] LAGACHERIE P., RABOTIN M., COLIN F. *et al.*, "Geo-MHYDAS: a landscape discretization tool for distributed hydrological modeling of cultivated areas", *Computers & Geosciences*, vol. 36, no. 8, pp. 1021–1032, 2010.

[MEJ 14] MEJÍA A., DALY E., ROSSEL F. *et al.*, "A stochastic model of streamflow for urbanized basins", *Water Resources Research*, vol. 50, no. 3, pp. 1984–2001, 2014.

[PAS 10] PASSALACQUA P., DO TRUNG T., FOUFOULA-GEORGIOU E. *et al.*, "A geometric framework for channel network extraction from lidar: nonlinear diffusion and geodesic paths", *Journal of Geophysical. Research*, vol. 115, no. F01002, 2010.

[ROD 08] RODRIGUEZ F., ANDRIEU H., MORENA F., "A distributed hydrological model for urbanized areas – model development and application to case studies", *Journal of Hydrology*, vol. 351, nos. 3–4, pp. 268–287, 2008.

[ROD 13] RODRIGUEZ F., BOCHER E., CHANCIBAULT K., "Terrain representation impact on periurban catchment morphological properties", *Journal of Hydrology*, vol. 485, pp. 54–67, 2013.

[RUS 02] RUSS J.C., *The Image Processing Handbook*, 4th edition, CRC Press, Boca Raton, 2002.

[SAN 16] SANGIREDDY H., STARK C.P., KLADZYK A. *et al.*, "GeoNet: An open source software for the automatic and objective extraction of channel heads, channel network, and channel morphology from high resolution topography data", *Environmental Modeling and Software*, vol. 83, pp. 58–73, 2016.

[SAN 17] SANZANA P., GIRONÁS J., BRAUD I. *et al.*, "A GIS-based urban and peri-urban landscape representation toolbox for hydrological distributed modeling", *Environmental Modelling & Softaware*, vol. 91, pp. 168–185, 2017.

[SHE 96] SHEWCHUCK J., "Triangle: Engineering a 2D Quality Mesh Generator and Delaunay Triangulator", in LIN M.C., MANOCHA D. (eds), *Applied Computational Geometry: Towards Geometric Engineering*, Berlin, Heidelberg, 1996.

[SMI 01] SMITH J., LYNN M., MORRISON J. *et al.*, "The Regional Hydrology of Extreme Floods in an Urbanizing drainage Basin", *Journal of Hydrometeorology*, vol. 3, pp. 267–282, 2001.

[WMO 08] WMO, Urban flood risk management – a tool for integrated flood management, associated programme on flood management, Technical report, 2008.

[XIA 07] XIAO Q., MCPHERSON E.G., SIMPSON J.R. *et al.*, "Hydrologic processes at the urban residential scale", *Hydrologic Processes*, vol. 21, pp. 2174–2188, 2007.

[ZEC 94] ZECH Y., SILLEN X., DEBOURCES C. *et al.*, "Rainfall runoff modelling of partly urbanized watersheds, Comparison between a distributed model using GIS and other models. Sensitivity analysis", *Water Science and Technology*, vol. 29, nos.1–2, pp. 163–170, 1994.

6

Mapping of Drought

6.1. Context

In semiarid regions, drought is a frequent phenomenon that causes serious problems for agricultural crops and that constitutes a real threat to food security. This problem is likely to greatly worsen in different regions of the world due to the effects of climate change and the strong growth in extreme phenomena. The different types of drought are:

– meteorological drought, which corresponds to a prolonged period of precipitation below the average;

– agricultural drought, which is a lack of water content for crops. This could occur even with normal rainfall, due to inadequate choices in agricultural practices;

– hydrological drought, which corresponds to a decrease in water reserves to a level below the average, in groundwater and other water reservoirs.

Various studies have focused on meteorological drought through the development of different types of drought indices. In general, these indices use *in situ* rainfall data. The indices most commonly used by the scientific community and managers are the Palmer Drought Severity Index (PDSI) [PAL 65], or the Standardized Precipitation Index (SPI) [MCK 95]. The calculation of PDSI is based on monthly series of precipitation and temperature. The SPI is based solely on precipitation data. In areas with very dense *in situ* meteorological stations, drought conditions can be interpolated with only small associated errors. For regions with

Chapter written by Mohammad EL HAJJ, Mehrez ZRIBI, Nicolas BAGHDADI and Michel LE PAGE.

few meteorological stations, statistical techniques such as the inverse distance weight method or the ordinary Kriging model may be considered. These methods are not sufficient to ensure precise quantification of drought in areas where *in situ* measurements are limited.

The problem is even more complicated in global regions where meteorological stations are totally absent. For this reason, in the last two decades, scientists have proposed various new approaches to estimating drought conditions, essentially based on the analysis of statistical anomalies derived from time series of optical satellite images [AMR 11, KOG 95, ZRI 16]. Optical remote sensing has demonstrated its great potential for monitoring vegetation dynamics because of its spatiotemporal coverage. The Normalized Difference Vegetation Index (NDVI) is widely used to meet this need. This index, which corresponds to an expression of the contrasting reflectance between the red and near infrared of a surface spectrum, is an easy-to-use quantity that can be linked to the green vegetation cover or abundance of vegetation.

The NDVI, which is sensitive to the presence of green vegetation, has been used for several regional and global applications in studies concerning the distribution and potential photosynthetic activity of vegetation. The statistical anomalies detected in the temporal evolution of this index could be directly linked to drought events.

In this chapter, we propose the use of a drought index based on a temporal series of NDVI MODIS images for the monitoring of drought in France during the last 15 years at a spatial resolution of 250 m. The goal is to show how to track and map drought using QGIS free software. The areas of agricultural, urban and forest regions will be provided for drought-affected zones in an example that corresponds to the period of the heatwave that affected Europe in the summer of 2003 (in the first half of August) [STE 12].

6.2. Satellite data

6.2.1. *MODIS products*

In this chapter, the MODIS[1] product containing the vegetation indices MOD13Q1 is used to map the drought. This product offers vegetation index images with a temporal resolution of 16 days and a spatial resolution of about 250 m ×

1 https://lpdaac.usgs.gov/dataset_discovery/modis/modis_products_table/mod13q1.

250 m. The MOD13Q1 data can be downloaded from the Reverb ECHO[2] website under the Hierarchical Data Format (HDF). Each MOD13Q1 image is composed of 12 bands. The projection system of the product MOD13Q1 is Sinusoidal.

In this chapter, we are only interested in bands 1 and 3 of the product MOD13Q1. The first band is the vegetation index (NDVI), and the third band gives information on the quality and reliability of this index. The quality band is composed of 16 bytes[3] (for example: 0000100001000100). We consider an NDVI pixel reliable if the value of the pixel in the quality band ends with 00.

To facilitate manipulation of the MOD13Q1 images, the NDVI band (band 1) of each MDO13Q1 image is separated into a single-band image. In this NDVI single-band image, we assign to unreliable pixels a No Data value.

6.2.2. *Land cover map*

A land cover map[4] with a spatial resolution of 30 m × 30 m is used to determine the areas of agricultural, urban and forest areas for regions affected by drought in 2003. This land cover map, produced by the Theia Center of Science (https://www.theia-land.fr/), is a thematic raster file with integer values between 11 and 222, where each value represents a type of land cover[5]. The projection system associated with the land cover map is the Lambert 93.

6.3. Drought index based on satellite NDVI data

Various indices for drought quantification have been proposed based on time series of optical images (AVHRR, SPOT-VGT, MODIS, etc.). In this chapter, we use the Vegetation Condition Index (VCI) proposed by Kogan *et al.* [KOG 95]. This index has been tested at different sites, revealing its strong potential for detecting and monitoring the drought. Kogan *et al.* [KOG 95], Seiler *et al.* [SEIL 98] and Quiring *et al.* [WHO 10] observed a high correlation between VCI and agricultural yield in different regions of the globe (Africa, America, Europe). Gitelson *et al.*

2 https://reverb.echo.nasa.gov/reverb/.

3 https://vip.arizona.edu/documents/MODIS/MODIS_VI_UsersGuide_June_2015_C6.pdf.

4 http://osr-cesbio.ups-tlse.fr/echangeswww/TheiaOSO/OCS_2014_CESBIO_L8.tif

5 http://osr-cesbio.ups-tlse.fr/~oso/ui-ol/2009-2011-v1/layer.html.

[GIT 98] showed that, when tested in six different sites in Kazakhstan, the VCI could account for 76% of the variations observed in cultivated areas. The VCI for a given date *t* that corresponds to the MODIS image synthesized on 16 days is defined by:

$$VCI_t = 100 \frac{(\mathrm{NDVI}_t - \mathrm{NDVI}_{\mathrm{min}})}{(\mathrm{NDVI}_{\mathrm{max}} - \mathrm{NDVI}_{\mathrm{min}})} \quad [6.1]$$

where NDVI_t is the MODIS NDVI at a given date *t* that corresponds to a MODIS image synthesized on 16 days, $\mathrm{NDVI}_{\mathrm{min}}$ is the NDVI multiyear minimum calculated using all the NDVI images available for the month containing the date *t* and $\mathrm{NDVI}_{\mathrm{max}}$ is the maximum NDVI calculated using all NDVI images available for the month containing date *t*. The VCI varies between 0 and 100. For vegetation areas, the VCI varies from a cover in a dry state (VCI = 0) to a maximum development state (VCI = 100). In this chapter, two classes of drought are identified [AMA 17]:

– extreme to moderate drought: VCI $\leq$ 20%;

– no drought: VCI > 20%.

6.4. Methodology

The steps for mapping the drought for the second half of August 2003 are as follows:

– step 1: calculate the minimum and maximum values of NDVI using all NDVI MODIS synthesized on 16 days available for August between 2000 and 2017;

– step 2: calculate the VCI drought index for a date just after the end of the 2003 heatwave period (the date that corresponds to the NDVI MODIS synthesized between August 13 and 29, 2003);

– step 3: change the value of the VCI pixels to 0 and 1: value 0 for pixels with a VCI $\leq$ 20% (extreme to moderate drought) and value 1 for pixels with a VCI > 20% (no drought);

– step 4: apply a majority filter on the VCI image coded in 0 and 1 (image from the previous step). This filter helps in highlighting the major areas affected by drought. The chosen size of the filter is 40 pixels × 40 pixels, which corresponds to an area of 10 km × 10 km;

– step 5: polygonize the filtered image (raster to vector) to delineate areas affected by drought;

– step 6: calculate, for each polygon that corresponds to a drought-affected zone, the areas of agricultural, urban and forest areas;

– step 7: using a map, show the spatial extent of drought zones and the area of agricultural, forest and urban regions within these zones.

6.4.1. *Preprocessing of MOD13Q1 images (step 1)*

Two scripts are developed to preprocess MODIS images (Figure 6.1). The first script is used to separate the NDVI band (band 1) from the product MOD13Q1, into a single-band image (tif format). The second serves to calculate the minimum and maximum NDVI values from the MODIS image time series (2000–2017) (Figure 6.1). These scripts are first added to QGIS. These two scripts are provided with this chapter's dataset.

QGIS functionality to add a script in QGIS:
- Processing> Toolbox> Scripts> Tools> Add script from file>...

QGIS functionality to launch the script from QGIS:
- Processing> Toolbox> Scripts> Tools> user scripts>...

6.4.2. *Delimitation of drought zones (steps 2–5)*

First, the drought index VCI is calculated for the period corresponding to the synthesis of the NDVI MODIS images between August 13 and 29, 2003. The pixels of the VCI image are then coded as 0 if the value of the VCI is less than or equal to 20% (extreme to moderate drought) and 1 if the VCI is greater than 20% (little or no drought). To obtain large homogeneous zones, a majority filter (size 40 × 40 pixels) is applied to the VCI image coded in 0 and 1. The filtered image is then polygonized (raster to vector) and finally the polygons representing areas that are not affected by drought are eliminated (Figure 6.1).

QGIS functionality to calculate the VCI and convert VCI values to 0 and 1:
- Raster> Raster Calculator> ...

QGIS functionality to apply a majority filter:
- Processing> Toolbox> SAGA> Raster filter> Majority filter> ...

QGIS functionality to polygonize a raster:
- Raster> conversion> polygonize> ...

QGIS functionality to remove polygons without drought:
- Right click on the vector layer> Open attribute table> Select features using an expression> Delete selected features> ...

6.4.3. *Calculate the area of agricultural, urban and forest zones affected by the drought (step 6)*

This last step (Figure 6.1) consists of calculating, for each polygon (zone) affected by the drought, the area of the three classes: agricultural, urban and forest zones (for each class, multiply the number of pixels by the area of a pixel). Finally, a visualization of the area of each class is presented in the form of a map.

In this chapter, the agricultural zone class includes the following classes of land cover map: summer crop (code 11), winter crop (code 12), grassland (code 211), orchard (code 221) and vine (code 222). The forest class contains the following classes of the land cover map: hardwood forest (code 31) and coniferous forest (code 32). Finally, the urban class contains the constructed classes (code 41) and diffracted constructions (code 42).

QGIS functionality to derive from the land use map three images respectively for "agricultural", "forests", and "urban" zones
- Raster> Raster calculator> ...

QGIS functionality to calculate the areas of the three classes: agricultural, forest and urban areas within each polygon identified as a drought-affected zone:
- Raster> Zonal Statistics> ...
- Right click on the vector layer> Open the attribute table> ...

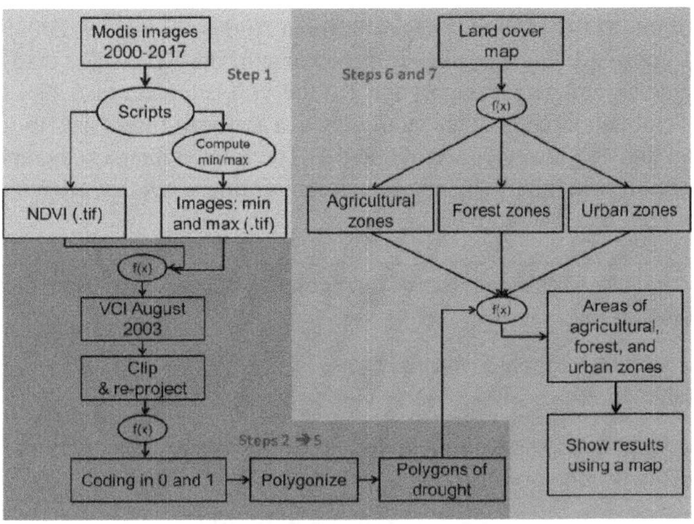

Figure 6.1. *Steps for drought mapping. For a color version of the figure, see www.iste.co.uk/baghdadi/qgis4.zip*

6.5. Implementation of the application via QGIS

This section contains the implementation of the QGIS functions for mapping drought.

6.5.1. *Download MODIS MOD13Q1 data*

MODIS data is downloaded from the Reverb ECHO website[6].

<table>
<tr>
<td>1. Download MODIS images</td>
<td>
➢ Select MODIS images to download:
• Go to the ECHO Reverb website: https://reverb.echo.nasa.gov/reverb/. In order to log in, it is necessary to first create a user account.
• Use the cursor to select the study area (France). Four MODIS tiles are needed to cover all of France. In this chapter, only the tile (h18) which covers almost all of France is used.
• In "Search Terms", write "MOD13Q1".
• In "START" and "END", specify the start (01/01/2000) and end (23/05/2017) dates of the MODIS acquisitions.
• In "Select Datasets", check the option "MODIS/Terra Vegetation Indices 16-Day L3 Global 250m SIN Grid V006".
• Click on "Search for Granules" to display the available images between 2000 and 2017.
• Click on the "ALL" button to select all available images between 2000 and 2017, then click on "View Items in Cart".

</td>
</tr>
</table>

6 https://reverb.echo.nasa.gov/reverb/.

- Finally, click on "Order", to receive by e-mail a link to download all the images.

Given the number of images selected, it is best to download them automatically one by one. One solution is to use the plugin "DownThemAll" of the Internet browser "Mozilla Firefox".

➢ Download selected images with the "DownThemAll" plugin

- Open "Mozilla Firefox" and install the "DownThemAll"[7] plugin.
- Using "Mozilla Firefox", open the link received by e-mail and right click ➔ "DownThemAll".

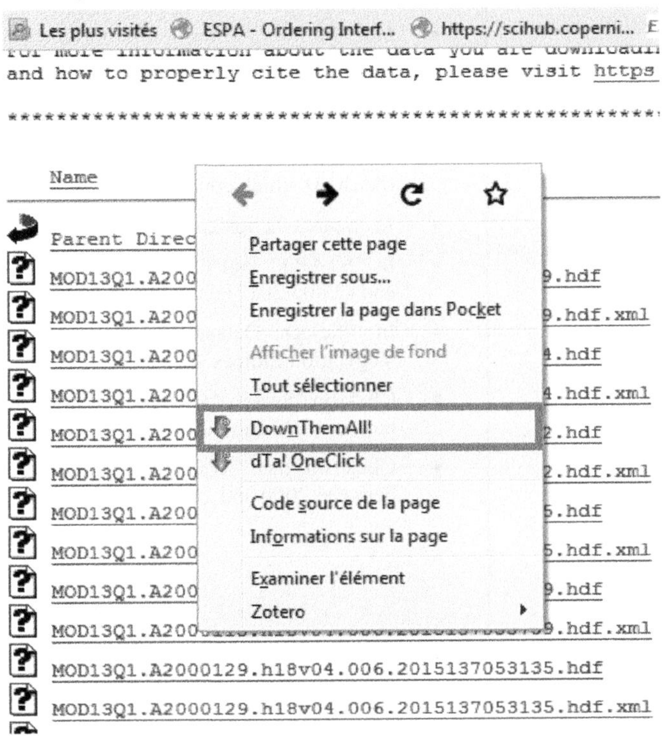

- In the window that appears, select the links to the images, and enter a directory in which to store the images.
- Finally, click on "Start" to start the download.

7 https://addons.mozilla.org/fr/firefox/addon/downthemall/.

	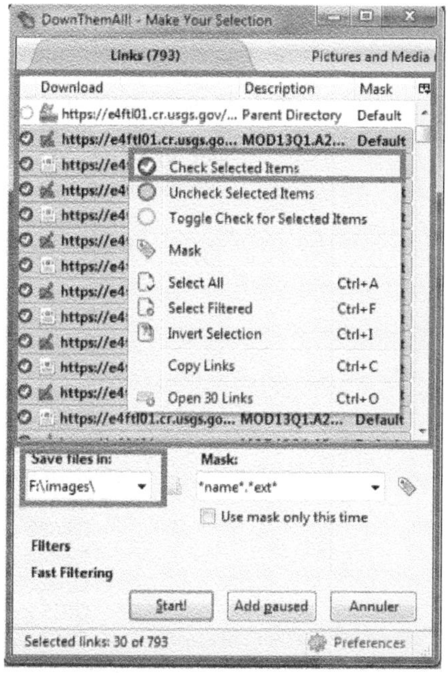

Table 6.1. *Download MODIS MOD13Q1data. For a color version of the table, see http://www.iste.co.uk/baghdadi/qgis4.zip*

6.5.2. *Preprocessing of MODIS MOD13Q1 data (step 1)*

Each of the downloaded MOD13Q1 images consists of 12 bands. However, to calculate the VCI we are only interested in the band NDVI (band number 1). To facilitate the manipulation of NDVI bands, a script in python language has been developed to separate the NDVI band from the MOD13Q1 product into a single-band image (".tif" format). This script assigns a **NoDATA** value to the NDVI pixels judged to be unreliable by the product quality band MOD13Q1 (band 3). The script is first added to QGIS.

1. Separate the NDVI band into a single-band image (".tif" format)	➢ Add the script to QGIS: • In the menu bar, click on "Processing" ➔ "Toolbox". • In the window that appears, click on "Scripts"➔ "Tools" ➔ "Add script from file" and select script "hdf_to_tif.py". The "hdf_to_tif.py" script is provided with the chapter database. 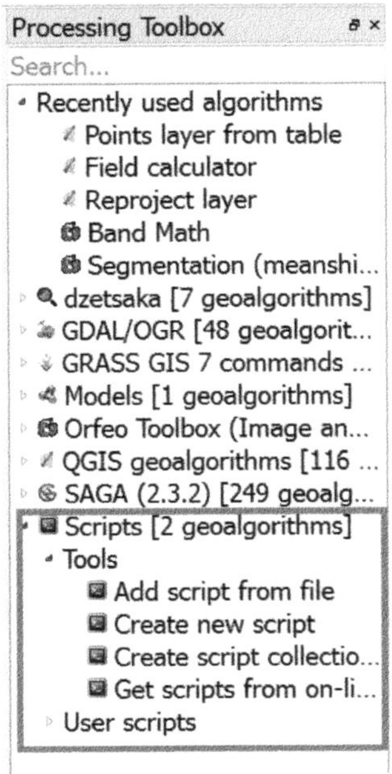 ➢ Launch the script from the QGIS interface: • In the menu bar, click on "Processing" ➔ "Toolbox". • Click on "User scripts" to display the script added in the previous step "hdf_to_tif.py". • Right click on the script "hdf_to_tif.py" ➔ "Edit script". • In the text window that appears, enter the name of the MODIS MOD13Q1 image directory, and save. In the example below, the MODIS images are located in the directory "J: / Drought / hdf / 16days / test /".

	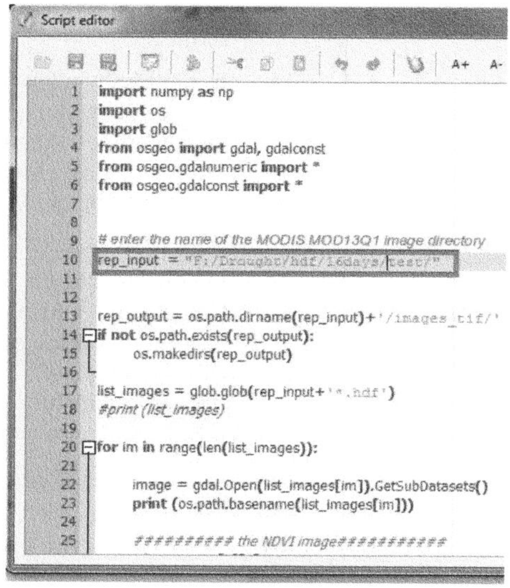 • Finally, to launch the script, right click on the script "hdf_to_tif.py" → "Run". NDVI mono-band images automatically save to a folder called "images_tif". This folder is created by the script in the same directory as the MODIS MOD13Q1 images.

Table 6.2. *Preprocessing of MODIS MOD13Q1 data. For a color version of the table, see www.iste.co.uk/baghdadi/qgis4.zip*

6.5.3. *Calculate VCI index (steps 1 and 2)*

First, two images containing, respectively, the minimum and maximum values of NDVI are derived from the MODIS time series images of August (between 2000 and 2017). A script in python language is developed to derive these two images. Then, to calculate the VCI, the NDVI image that corresponds to the synthesis of the MODIS NDVI images between August 13 and 29, 2003 is used.

1. Calculate the minimum and maximum values of the NDVI	To calculate at each pixel the minimum and maximum values of the NDVI index: • Put all MODIS MOD13Q1 images acquired in August (2000-2017) into a folder. • Add the script "image_min_max.py" to QGIS (see section 6.5.2). • Click on "User scripts" to display the script added in the previous step "image_min_max.py".

<table>
<tr>
<td></td>
<td>

- Right click on the script "image_min_max.py" ➔ "Edit script".
- In the text window that appears, enter the name of the directory containing the MODIS MOD13Q1 images for the month of August (2000–2017), and save. In the example below, MODIS images of August are located in the directory: J:/Sechresse/hdf/16days/images_august/.
- Finally, to launch the script, right click on the script "image_min_max.py" ➔ "Execute". The minimum and maximum values of the NDVI are stored in a folder called "min_max". This folder is created by the script in the same directory containing the MODIS MOD13Q1 images for the month of August (2000–2017).

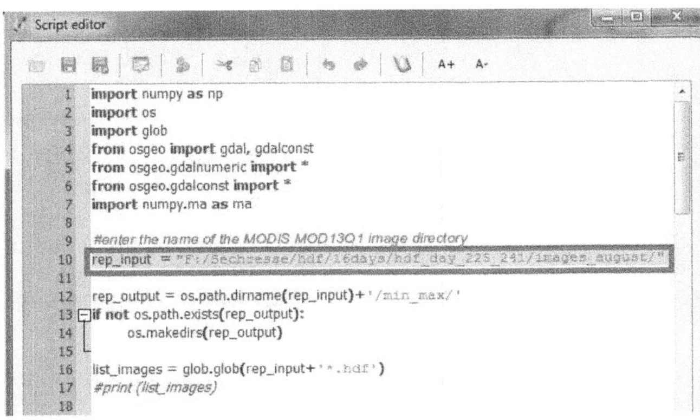

</td>
</tr>
<tr>
<td>2. Calculate the VCI</td>
<td>

➢ To calculate the VCI index :

- Open QGIS and import the images : "MOD13Q1.A2003241.h18v04.006.2015153114148.tif", " min_2000_2017.tif", " max_2000_2017.tif".
- In the menu bar, click on "raster" ➔ "raster calculator".
- In the window that appears, type the formula below

100 * (("MOD13Q1.A2003241.h18v04.006.2015153114148@1" - "min_2000_2017@1")/("max_2000_2017@1"- "min_2000_2017@1"))

- In "output layer", name the image as "VCI_20030829.tif".
- Finally, click on "OK". Once processing is complete, the output image appears in the "layers" and "display" parts of the QGIS interface.

</td>
</tr>
</table>

<table>
<tr><td></td><td>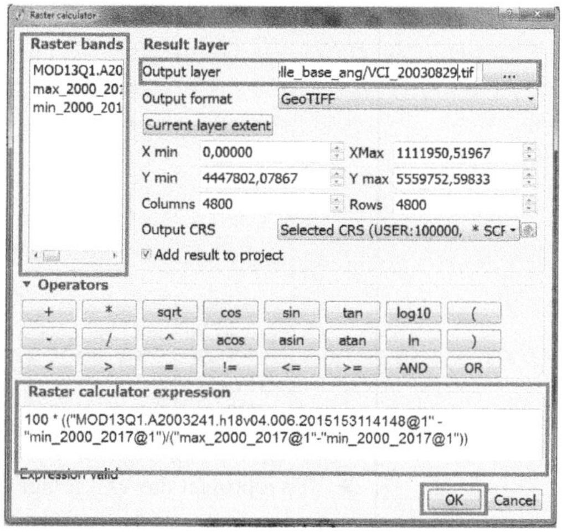
</td></tr>
<tr><td>3. Clip and reproject the VCI image</td><td>
The VCI image crosses the border of France on the east side and is encoded in Float 32-bit. To facilitate further processing, the VCI image is clipped to fit the spatial extent of France, converted into a 16-bit integer, and reprojected to Lambert 93.
➢ To clip the VCI image and convert it into a 16-bit integer:
<ul>
<li>Import into QGIS the image "VCI_20030829.tif" and the shape file delimiting the "France.shp" study area. This layer is provided with the chapter database.</li>
<li>In the menu bar, click on "raster" → "extraction" → "clipper".</li>
<li>In the window that appears, select the image "VCI_20030829.tif" in "Input file".</li>
<li>In "Output file", give the output image the following name: "VCI_20030829_clip.tif".</li>
<li>Enable the "No data value" option and enter –32768.</li>
<li>Check the "Mask layer" option, and enter the shape file "France.shp" provided with the database.</li>
<li>Activate the option "Crop the extent of the target dataset to the extent of the cutline".</li>
<li>Click on "Edit" and add "-ot Int16" in the text area just after "GTiff" to get the output image as a 16-bit integer.</li>
<li>Finally, click on "OK". Once the processing is complete, the image cut out and converted into 16 bits appears in the "layers" and "display" part of the QGIS interface.</li>
</ul>
</td></tr>
</table>

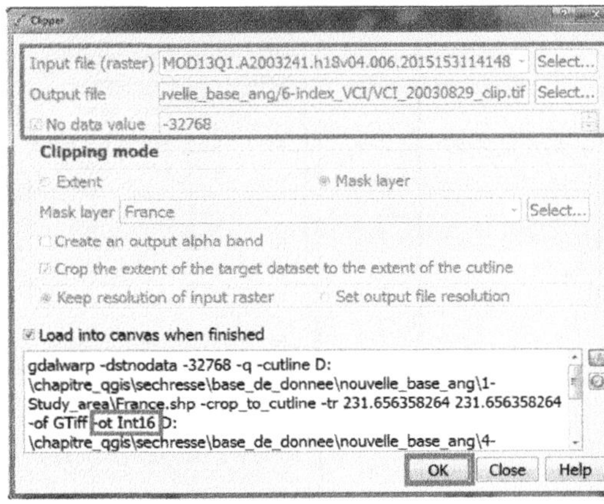

➤ To reproject the VCI image:

- In the menu bar, click on "raster" ➔ "projections" ➔ "warp (reproject)".
- In the window that appears, select the image "VCI_20030829_clip.tif" in "Input file".
- In "Output file", enter "VCI_20030829_clip_rgf93.tif".
- Activate the "Target SRS" option, and choose the projection "RGF93/Lambert-93 (EPSG: 2154)".
- Finally, click on "OK". Once the processing is complete, the output image appears in the sections "layers" and "display" of the QGIS interface.

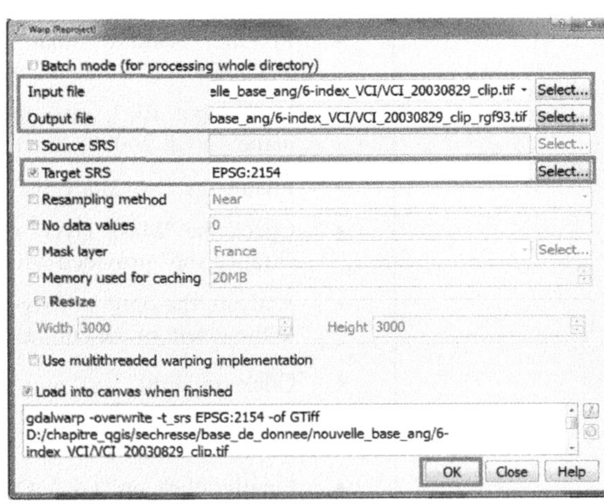

Table 6.3. *Calculation of VCI. For a color version of the table, see www.iste.co.uk/baghdadi/qgis4.zip*

6.5.4. *Delimitation of drought zones (steps 2–5)*

To delineate the drought zones, first the pixels of the VCI image are coded at 0 if the value of the VCI is less than or equal to 20% (extreme to moderate drought) and 1 if the VCI is greater than 20% (no drought). Then, to obtain large homogeneous zones, a majority filter is applied to the VCI image coded at 0 and 1 (size 40 × 40 pixels). The filtered image is then polygonized (raster to polygon) to delimit the different homogeneous regions. Finally, polygons corresponding to areas not affected by drought are eliminated.

<table>
<tr>
<td>1. Coding the VCI image at 0 and 1</td>
<td>

➢ To encode the VCI image at 0 and 1:

- Import the VCI image: "VCI_20030829_clip_rgf93.tif".
- In the menu bar, click on "raster" ➔ "raster calculator".
- In the window that appears, type the formula below. With this formula, the pixels of the VCI image with values less than or equal to 20% take the value 0 in the output image, while the pixels with values greater than 20% take the value 1: ("VCI_20030829_clip_rgf93 @ 1"> 20)
- In "Output layer", name the image as follows: "VCI_20030829_c01.tif".
- Finally, click on "OK". After processing, the output image appears in the "layers" and "display" part of the QGIS interface.

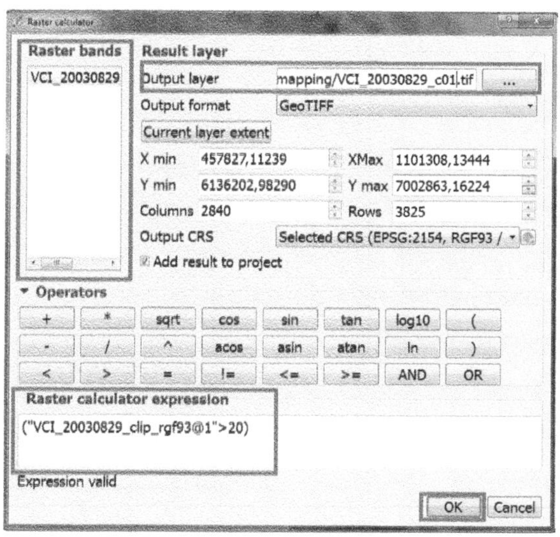

</td>
</tr>
</table>

➢ The "raster calculator" function produces the "VCI_20030829_c01.tif" image with a 32-bit floating-point format. To facilitate further processing, it is important to convert the image "VCI_20030829_c01.tif" into a 16-bit integer:

- Import the VCI image "VCI_20030829_c01.tif" encoded at 0 and 1.
- In the menu bar, click on "raster" → "conversion" → "translate".
- In the window that appears, select in "Input layer" the image "VCI_20030829_c01.tif".
- In "Output file" enter "VCI_20030829_c01_int.tif".
- Enable the "no data" option and enter -32768.
- Click on "Edit" and add "-ot Int16" to the text area just after "GTiff".
- Finally, click on "OK". After processing, the output image appears in the "layer" and "display" part of the QGIS interface.

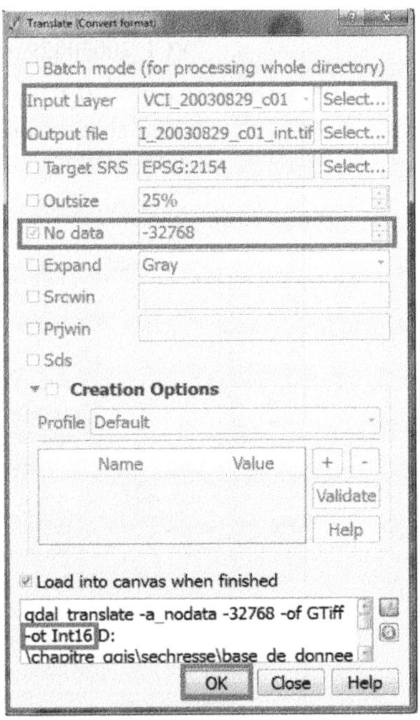

2. Apply a majority filter	➢ To apply the majority filter (size 40 × 40 pixels): • Import the VCI image "VCI_20030829_c01_int.tif". • In the menu bar, click on "Processing" ➔ "Toolbox" ➔ "SAGA" ➔ "raster filter" ➔ "majority filter". • In the window that appears, select the image "VCI_20030829_c01_int.tif" in "Grid". • In "Search Mode", select "square". • In "Radius", enter 40 (the unit is pixels). • In "Filtered Grid", enter the output file name "VCI_20030829_c01_int_maj.tif". • Finally, click on "OK". After processing, the output image appears in the "layer" and "display" part of the QGIS interface. 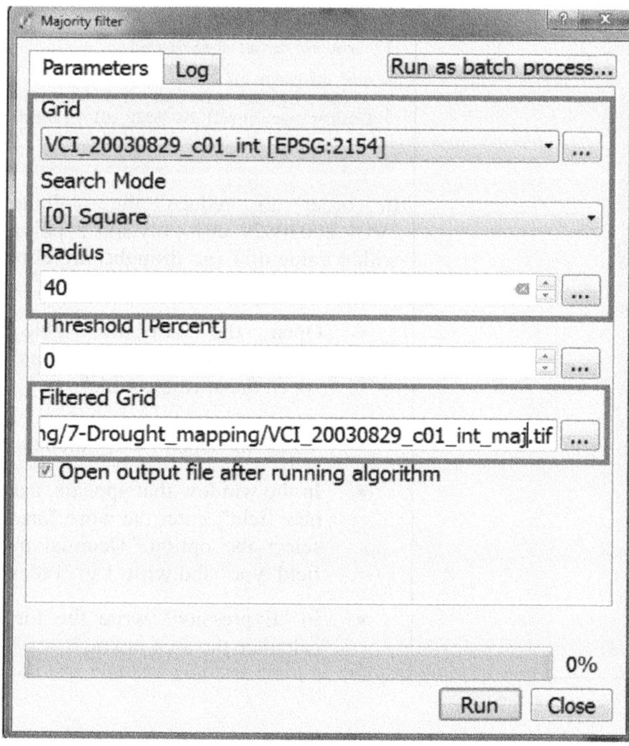

<table>
<tr>
<td>3. Polygonize the VCI image</td>
<td>➢ To polygonize the VCI image:
• Import the image "VCI_20030829_c01_int_maj.tif".
• In the menu bar, click on "raster" ➔ "conversion" ➔ "polygonize (raster to vector)".
• In "Output polygon file (shp)" enter the name: "VCI_20030829_c01_poly.shp".
• Finally, click on "OK". After processing, the output polygon file appears in the "layer" and "display" part of the QGIS interface.
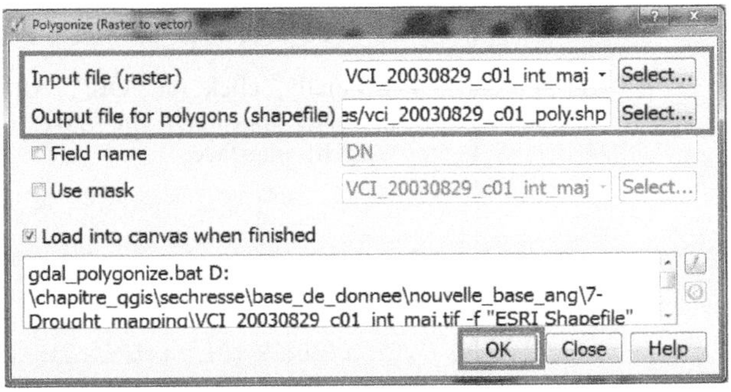

In order to reduce the volume of the vector file "VCI_20030829_c01_poly.shp", polygons that delimit pixels with a value of 1 (no drought) are eliminated. In addition, small polygons (<100 pixels or 6250000 m^2) are also deleted:
• Open the attribute table of the vector file "VCI_20030829_c01_poly.shp": right click on it ➔ "open the attribute table".
• Click on "Open Field Calculator" .
• In the window that appears, activate the option "Create a new field", enter the word "area" in "Output field name", select the option "Decimal number (real)" in "Output field type" and write 1 in "Precision".
• In "Expression" write the formula "$area/1000000" to calculate the area in km^2.</td>
</tr>
</table>

<table>
<tr>
<td></td>
<td>

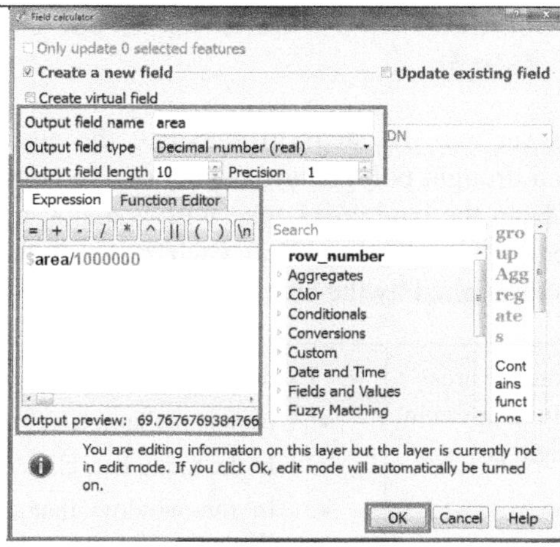

- Activate Edit mode .
- Click on "Select features using an expression" and write: "DN"=1 or "area" <6.25.
- Click on 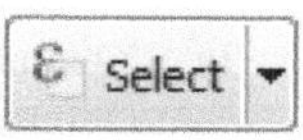, then on "Delete selected features" .
- Deactivate Edit mode .

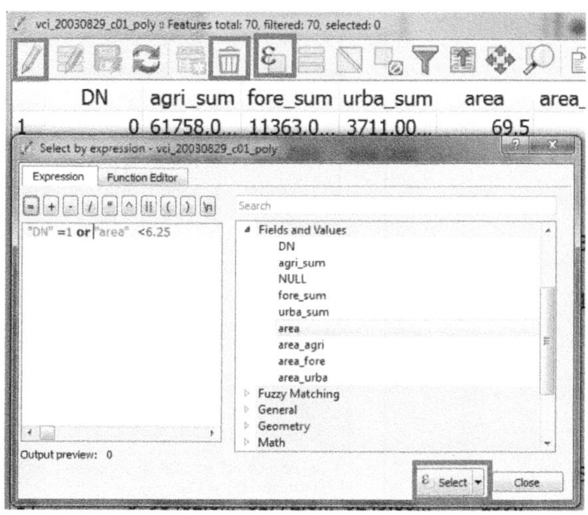

</td>
</tr>
</table>

Table 6.4. *Delimitation of drought zones. For a color version of the table, see www.iste.co.uk/baghdadi/qgis4.zip*

6.5.5. *Calculation of agricultural, forest and urban areas affected by drought (step 6)*

In this section, the steps to calculate the agricultural, forest and urban areas within each drought polygon are explained. To compute the areas, each class is first extracted from the land cover map with a spatial resolution of 30 m × 30 m. Then, the number of its pixels within each drought polygon is counted. Finally, this number is multiplied by the area of each pixel (900 m^2, pixel size 30 m × 30 m).

<table>
<tr>
<td>1. Calculate the areas of agricultural areas in each drought polygon</td>
<td>

➢ Extracting agricultural areas:

- Import the land cover map "landcover.tif".
- In the menu bar, click on "raster" ➔ "raster calculator".
- In the window that appears, type the formula below. With this formula, the pixels of the land cover image with values equal to 11 (summer crop), 12 (winter crop), 211 (permanent grassland) and 222 (vine) take the value 1 in the output image, while the other pixels are set to 0.

("landcover@1" = 11 OR "landcover@1" = 12 OR "landcover@1" = 211 OR "landcover@1" = 222)

- In "Output layer", name the image as follows: "agricultural_areas.tif".
- Finally, click on "OK". After processing, the output image appears in the "layer" and "display" part of the QGIS interface.

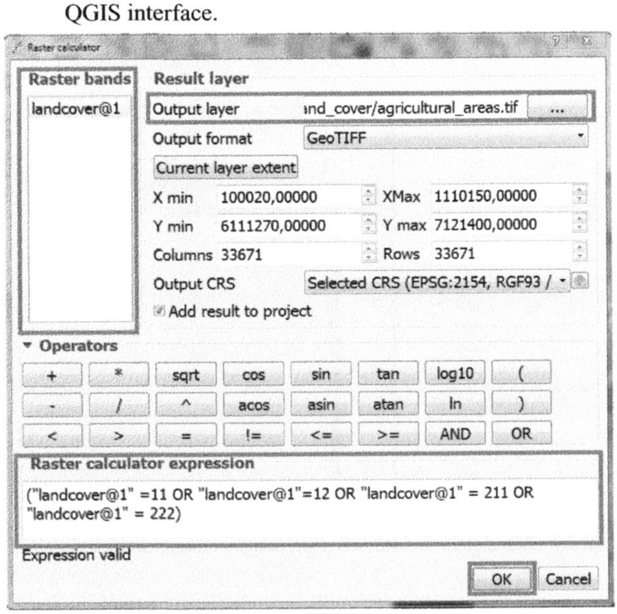

</td>
</tr>
</table>

- ➢ Calculate the number of pixels of agricultural areas within each drought polygon:
  - Import the shape file "VCI_20030829_c01_poly.shp" and the image "agricultural_areas.tif".
  - In the menu bar, click on "raster" ➔ "zonal statistics" ➔ "zonal statistics".
  - In "Raster layer" select the image "agricultural_areas.tif".
  - In the "polygon layer containing the zones" select the vector layer "VCI_20030829_c01_poly.shp".
  - In "Output column prefix" type "agri_".
  - In "Statistics to calculate" check only the option "Sum".
  - Click on "OK". After processing, statistics (number of pixels) are saved in the attribute table.

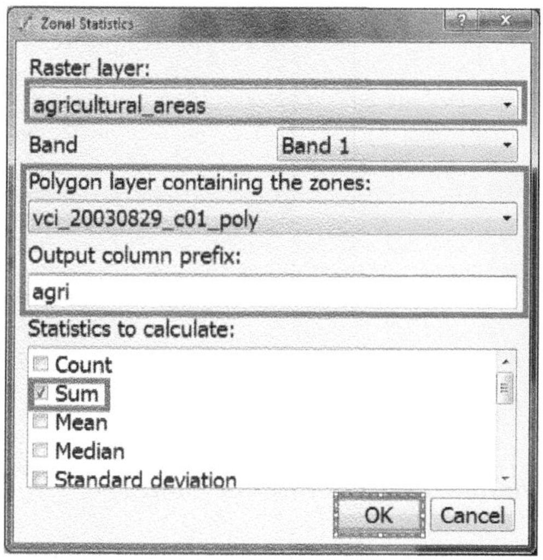

  - Repeat the above tasks in order to calculate the number of pixels in forest and urban areas.
- ➢ Calculate the area of agricultural, forest and urban regions affected by drought
  - Import the shape file "VCI_20030829_c01_poly.shp".
  - Open attribute table: right click ➔ "Open attribute table".
  - Click on "Open Field Calculator" .
  - In the window that appears, activate the option "Create a new field", enter the word "area_agri" in "Output field name", select the option "Decimal number (real)" in "Output field type" and write 1 in "Precision".

<table>
<tr>
<td></td>
<td>

- In "Fields and Values", double click on the column containing the number of pixels of the "agri_sum" agricultural areas. The name of the column appears in the "Expression" part. In the "Expression" part, complete the equation as follows: "agri_sum" * 30 * 30/1000000 (30 m is the resolution of the image occupation of the soils; it is divided by 1,000,000 to gain the result in km²).
- Repeat the above steps in order to calculate the area of forest and urban regions.

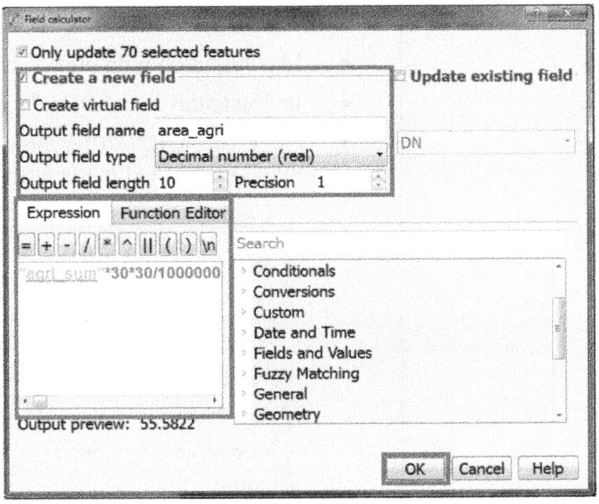

</td>
</tr>
</table>

Table 6.5. *Calculation of agricultural, forest and urban areas affected by drought. For a color version of the table, see www.iste.co.uk/baghdadi/qgis4.zip*

6.5.6. *Visualization of results (step 7)*

In this section, the results were presented using a map. For this map, only drought zones with area higher than 1,000 km^2 are considered.

<table>
<tr>
<td>1. View the results with a map</td>
<td>

➤ To show with a map the spatial extent of drought zones and the area of agricultural, forest and urban regions within these zones.

- Import the vector layer that contains the regions affected by drought "VCI_20030829_c01_poly.shp".
- Right click ➔ "Properties".
- In "Style", select the "Categorized" option, write "$id" in "Column" and click "Classify".

</td>
</tr>
</table>

<table>
<tr>
<td></td>
<td>
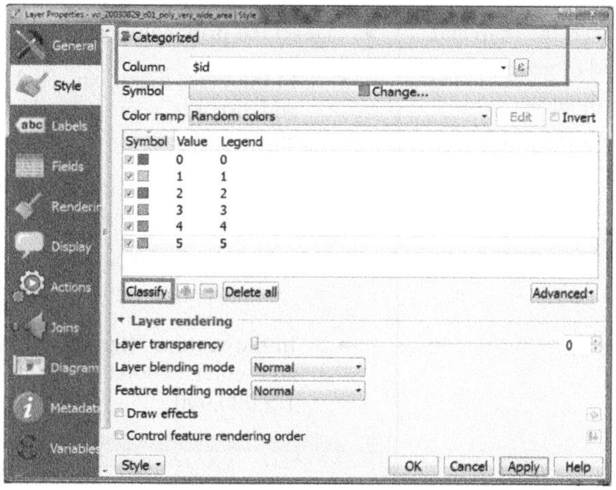

<ul>
<li>In "Diagrams", select the option "Pie chart".</li>
<li>In "Attributes", add the columns containing the areas of the agriculture "area_agri", forest "area_fore", and urban "area_urba".</li>
<li>Finally, click on "OK".</li>
</ul>

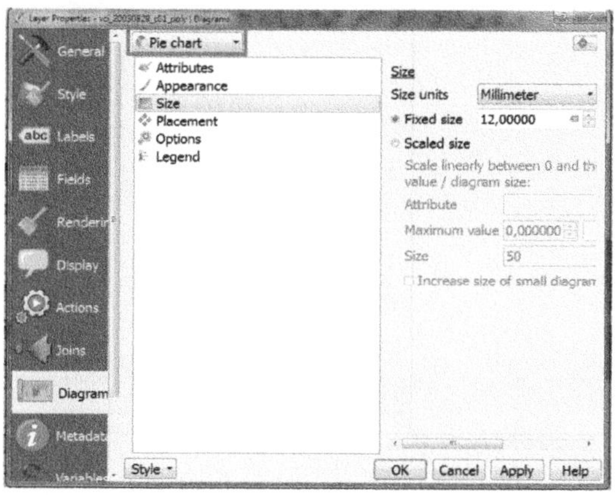

</td>
</tr>
<tr>
<td>2. Create a new print composer</td>
<td>
➢ To create a new print composer:
<ul>
<li>Import the shape files "VCI_20030829_c01_poly.shp" and "France.shp".</li>
<li>In the main menu, click on "Project" ➔ "new Print composer", or type: "Ctrl + P".</li>
<li>In the window that appears, give a name to the print composer, for example "drought", and click "OK".</li>
</ul>
</td>
</tr>
</table>

<table>
<tr>
<td>3. Insert the map and a grid of geographical coordinates</td>
<td>

➢ To set the print composer size and scale:

- In the main menu bar, click on "Layout" ➔ "Add Map", and use the mouse to draw a rectangular area on the existing sheet.
- In the menu bar, click on "Layout" ➔ "move item" and select the rectangular area containing the image.
- In the section "Composition", configure the sheet type as "A4 (210 mm × 297 mm)".
- In the section "Item properties", go to "Main properties" and enter 5,000,000 in "Scale" to define a scale of 1/5,000,000.

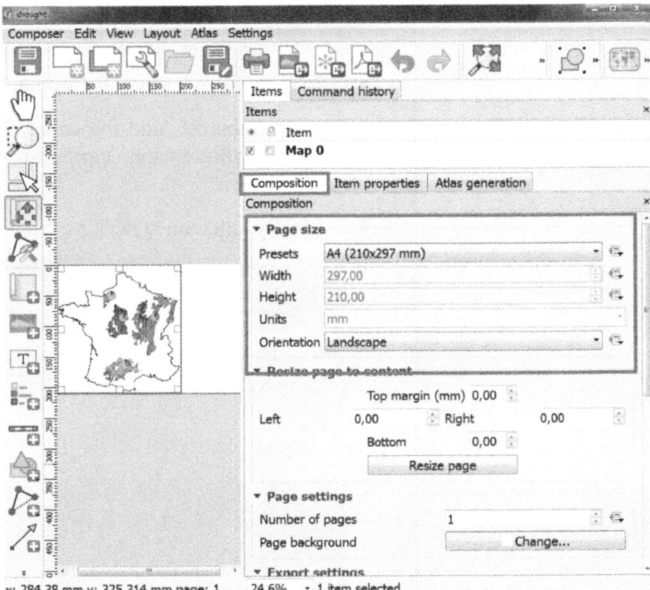

</td>
</tr>
</table>

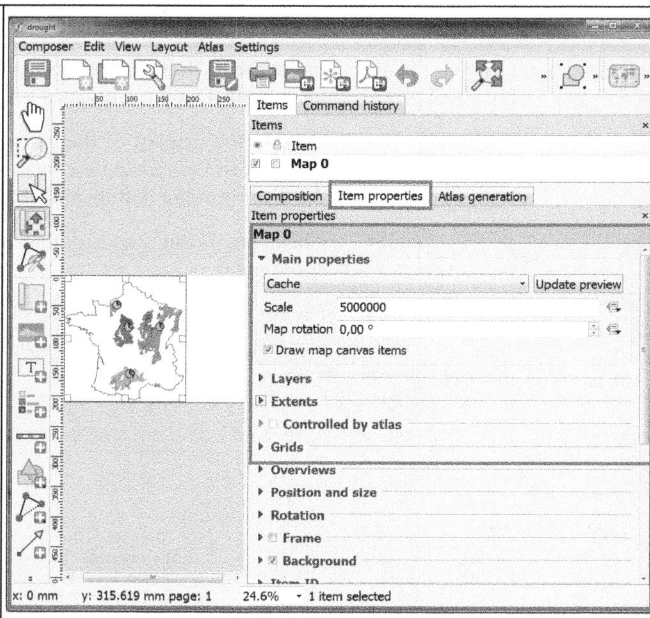

➤ To add a coordinate grid:

- In the section "Item properties", go to "Grids" and click on "+" to add a coordinate grid. In addition, define the grid interval as 200 km along the X-axis and the Y-axis.

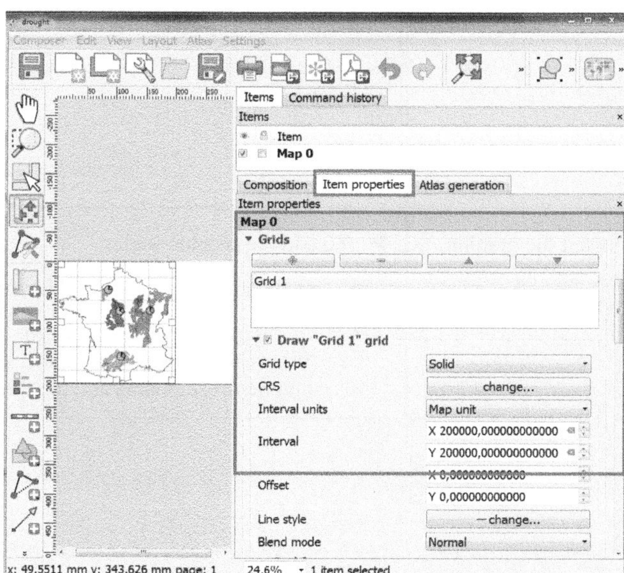

	➢ To display the Cartesian coordinates: • In the menu "Item properties", go to "Grids" and activate the option "Draw coordinates". • Disable the display of the coordinates at the top and at the right of the rectangle containing the image, and keep the display at the bottom and at the left. • Click on "Font" and define font and font size of coordinates. • On the field "Coordinate Precision" enter 0. 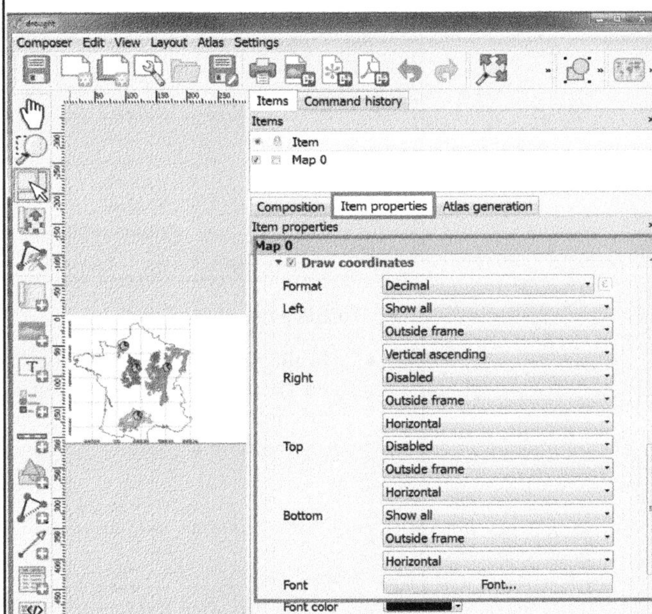
4. Insert the graphic scale and the arrow of the North	➢ To add the graphic scale: • In the menu bar, click on "Layout" ➔ "add scalebar". • In "Item properties", go to "Units", "Segments" and "Fonts and colors" to define, respectively, the unit, the number of segments and the font of the graphic scale.

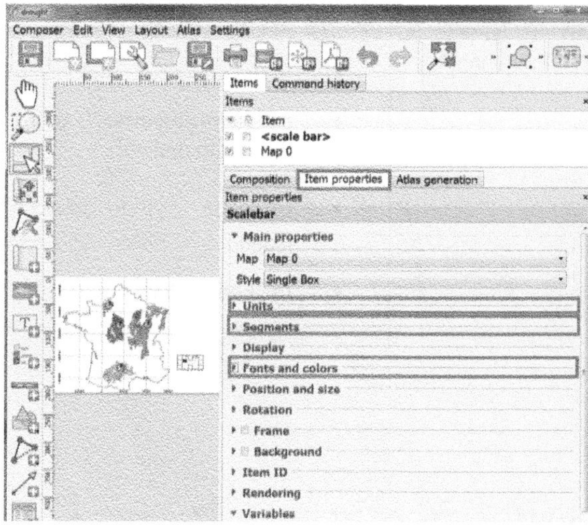

➢ To add the North arrow:

- Download an image of the North arrow from the Internet[8], then save this image on the computer.
- In the menu bar, click on "Layout" ➔ "Add Image" and draw a rectangular area.
- Select the rectangular area drawn. Then, in the field "Item properties", go to "Main properties" and in "Image source", enter the photo of the North arrow downloaded from the website.

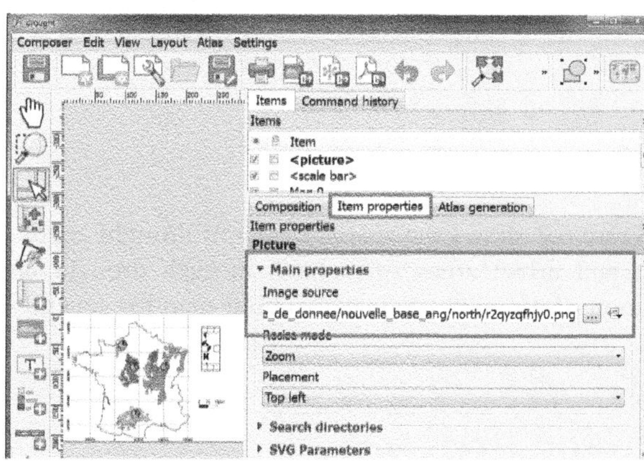

8 https://goo.gl/images/mns0lM.

5. Insert the legend and print the map	➤ To add the legend: • In the menu bar, click on "Layout" ➔ "Add legend". • In "Item properties", go to "Legend items" and disable the "Auto update". • Click on  to delete all parts of the uninteresting legend. • In "Font", change the fonts of the legend. 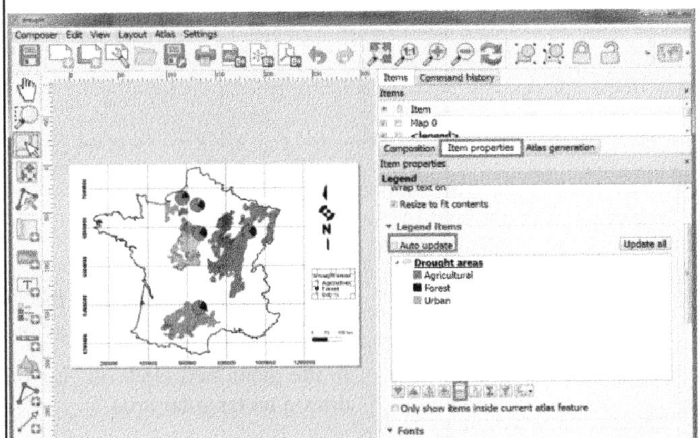 ➤ To print the map: in the menu bar, click "Composer" ➔ "Print composer" ➔ "Export as PDF".

Table 6.6. *Visualization of results. For a color version of the table, see www.iste.co.uk/baghdadi/qgis4.zip*

6.6. Drought map

Figure 6.2 shows the spatial extent of drought zones and the part of agricultural, forest and urban areas (pie chart) in each drought area. To study the temporal evolution of drought, produce a similar map for several dates and compare between them.

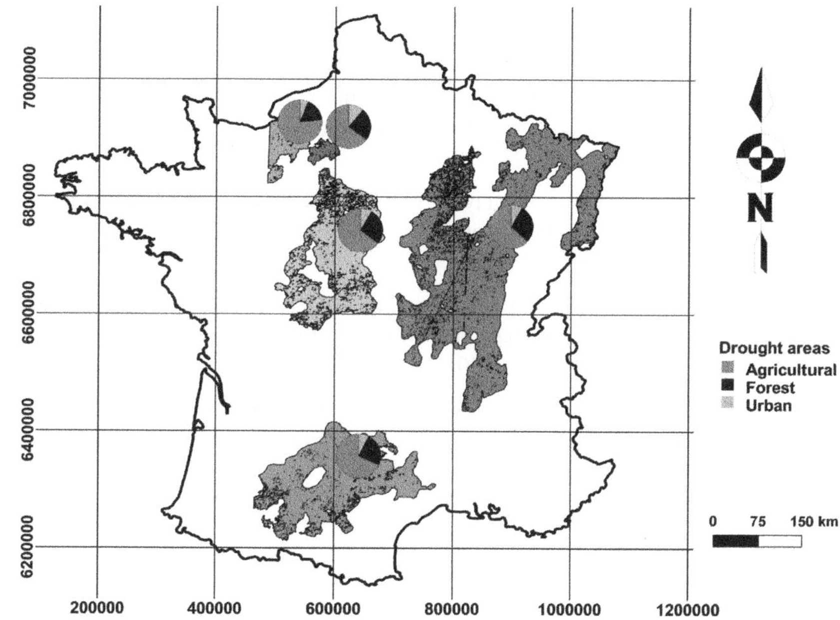

Figure 6.2. *Map of drought (August 2003). The coordinates are in Lambert 93. The drought areas have been colored to clearly show the relation to the pie chart (a pie chart was established for each of these areas). For a color version of the table, see www.iste.co.uk/baghdadi/qgis4.zip*

6.7. Bibliography

[AMA 17] AMALO L.F., HIDAYAT R., "Comparison between remote-sensing-based drought indices in East Java", *IOP Conference Series: Earth and Environmental Science*, Bogor, Indonesia, p. 012009, 2017.

[AMR 11] AMRI R., ZRIBI M., DUCHEMIN B. *et al.*, "Analysis of vegetation behaviour in a semi-arid region, using SPOT-VEGETATION NDVI data", *Remote Sensing*, vol. 3, pp. 2568–2590, 2011.

[GIT 98] GITELSON A.A., KOGAN F., ZAKARIN E. *et al.*, "Using AVHRR data for quantitative estimation of vegetation conditions: Calibration and validation", *Advances in Space Research*, vol. 22, pp. 673–676, 1998.

[KOG 95] KOGAN F.N., "Application of vegetation index and brightness temperature for drought detection", *Advances in Space Research*, vol. 15, pp. 91–100, 1995.

[MCK 95] MCKEE T.B., DOESKEN N.J., KLEIST J., "Drought monitoring with multiple time scales", *Proceedings of the 9th Conference on Applied Climatology*, American Meteorological Society Dallas, Boston, MA, pp. 233–236, 1995.

[PAL 65] PALMER W.C., Meteorological Drought; US Department of Commerce, Weather Bureau Washington, DC, vol. 30, 1965.

[QUI 10] QUIRING S.M., GANESH S., "Evaluating the utility of the Vegetation Condition Index (VCI) for monitoring meteorological drought in Texas", *Agricultural and Forest Meteorology*, vol. 150, pp. 330–339, 2010.

[SEI 98] SEILER R.A., KOGAN F., SULLIVAN J., "AVHRR-based vegetation and temperature condition indices for drought detection in Argentina", *Advances in Space Research*, vol. 21, pp. 481–484, 1998.

[STE 12] STÉFANON M., DROBINSKI P., D'ANDREA F. *et al.*, "Effects of interactive vegetation phenology on the 2003 summer heat waves", *Journal of Geophysical Research: Atmospheres*, vol. 117, pp. D24103, 2012.

[ZRI 16] ZRIBI M., DRIDI G., AMRI R. *et al.*, "Analysis of the effects of drought on vegetation cover in a Mediterranean region through the use of SPOT-VGT and TERRA-MODIS long time series", *Remote Sensing*, vol. 8, p. 992, 2016.

7

A Spatial Sampling Design Based on Landscape Metrics for Pest Regulation: The Millet Head Miner Case Study in the Bambey Area, Senegal

7.1. Definition and context

The objective of this study is to develop a spatial sampling plan to measure the influence of landscape elements on the natural regulation of the Millet Head Miner (MHM) *Heliocheilus albipunctella* De Joannis (see [AJA 80, GUE 82, NDO 79]). Causing important yield losses ranging from 40 to 85%, the MHM constitutes a threat to food security in West Africa (see [GAH 86, KRA 95, YOU 98]). Characterized by small-scale farming with low income, the use of insecticides is not suitable for Senegal because of their prohibitive costs, low availability, as well as the lack of training in the use of chemicals [PAY 11]. Thereby, promoting natural regulation of pests by the use of plant biodiversity appears to be a good compromise that could improve not only yields, but also food diversity and, in fact, ensure food security and health for local populations.

Numerous studies show that increasing plant diversity could limit pest pressure in crops. Among the agroecological practices based on biodiversity conservation, the integration of seminatural landscape elements such as hedgerows, trees, thickets, grass strips and ditches can help to increase biological control of pests by offering suitable habitats for food or oviposition. These landscape elements are generally resource areas for beneficial insects or other pest predators, thus reducing the need for pesticide application. They could also have a repellent effect on pests. In order to

Chapter written by Valérie SOTI.

exploit landscape elements for pest control, it is necessary to have a good knowledge of local natural habitats and their impact on pests and their natural enemies [BIA 06, LAN 00].

Integrated pest management for crop protection is mostly limited to push pull strategies by the use of allelopathic plants at the plot scale. However, the insects' (pests/natural enemies) spatial and temporal dynamics exceed the boundaries of the crop plot. Therefore, understanding the ecological complexity at the agroecosytem level involves replacing the cultivated plot in its environment.

In order to study the interactions between distribution of landscape elements and ecological processes over time, landscape ecology methodology is advised because it combines both the spatial and the functional approaches. For this science, the landscape has a spatial heterogeneity composed of a mosaic of patches varying in terms of composition and structure (fragmentation/corridor/homogeneity, etc.) which influence the presence, the abundance and the dispersal of the species within it.

According to this approach, the objective of this chapter is to develop a sampling strategy to study the effect of landscape composition on the MHM in a square region 20 km × 20 km in size centered on the Dangalma village in the North-Est of Thies. The study area located in the great groundnut basin in Senegal is characterized by an agroforestry system dominated by *Faidherbia albida* centuries-old trees in association with millet (*Pennisetum glaucum*) and peanut (*Arachis hypogaea* L.), which are the main cultivated crops.

This sampling plan is based on two landscape variables considered to be levers likely to play an important role in the natural regulation of the MHM. The first variable concerns the natural vegetation, mainly composed of isolated trees whose presence and density is supposed to favor the natural regulation of the MHM. Indeed, natural vegetation can provide food and alternative prey as well as shelter for natural enemies [OTI 11]. The second concerns the millet crop density, supposedly influencing the MHM incidence, as a "source area" for the MHM, which feeds on the millet panicle during its development cycle [BIA 06].

To validate these two hypotheses, it was considered fundamental to propose a sampling design along a relevant gradient of landscape patterns with a manageable number of sites that are representative of the full landscape heterogeneity.

The two landscape variables were calculated from a detailed map derived from a Pleiades satellite image of very high spatial resolution (0.4 m resolution). The methodology was developed with the free software QGIS and does not require any special skill in GIS to be applied.

7.2. The spatial sampling methodology

The proposed sampling protocol is based on landscape data derived from a Pleiades satellite image acquired on January 24, 2013, with a ground resolution of 0.5 × 0.5 m in panchromatic mode, and 2 × 2 m in multispectral mode with blue (B), green (G), red (R) and near infrared (NIR) bands. Three major steps were needed for its implementation (Figure 7.1):

– the quantification of the two landscape metrics was carried out using a land cover map of the study area composed of the following eight classes: millet, groundnut, market gardening, fallow land, vegetation trees, habitats, roads and bare soils [BIL 13];

– the combination of the two landscape variables;

– and finally, the integration of the selected sampling sites in a GPS for a field validation mission.

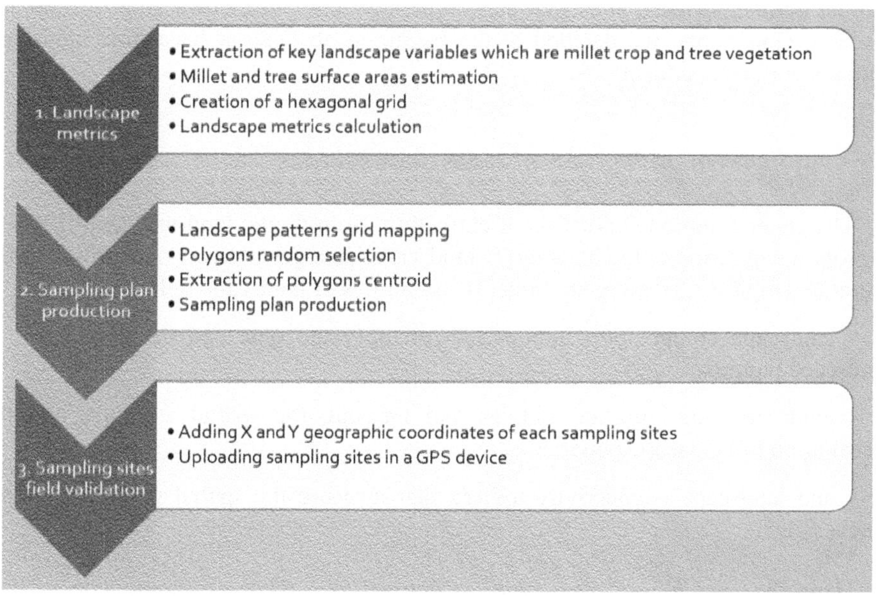

Figure 7.1. *The spatial sampling design processing steps*

Scientific and material limitations have led us to make some compromises for the sampling plan in terms of the number of sites and their replications.

The first constraint was to define a sufficient distance between the sampling sites to avoid any possible spatial correlation of the data among site observations. This separation distance generally coincides with the knowledge of the maximum dispersal capacity of the targeted insects. Concerning the MHM in Senegal, we know that females lay around 400 eggs on the panicle apex. After hatching, larvae start feeding on the millet head to complete their development over 23–29 days. Then, larvae pupate in the soil and diapause throughout the dry season from November to July. Thus, the MHM displacement distance is generally very short or null if the crop for next year is still millet. On the other hand, natural enemies that often use trees as a refugee host will tend to travel farther distances to attack pest insects. Based on the knowledge of MHM bioecology, we fixed the maximum displacement distance at a 1 km radius from the target millet plot, which allowed us to keep sample sites independent from each other to a certain degree.

The second constraint was to eliminate any edge effects by taking care to choose sites with a radius of 1 km located entirely within the study area. Finally, given the limited number of field sites that can be deployed, 45 in totality, due to time and resource constraints, we decided at the beginning to keep at least nine landscape patterns with five replications.

7.2.1. *Step 1: quantification of landscape metrics*

The general approach consists of extracting, from digital land-use data, the most appropriate descriptors to characterize landscapes in a generic way. According to the literature (Figure 7.2), there are three main families of landscape indices:

– landscape composition indices that quantify the type and the abundance of landscape patterns;

– landscape configuration indices that measure the spatial structure and the distribution of landscape patterns;

– and landscape connectivity indices that measure the spatial relation between patches.

Here, the "*patch*" or "class type" is defined as a relatively homogeneous area that differs from its surroundings. Patches are the basic unit of the landscape that change and fluctuate. *Matrix* is the "background ecological system" of a landscape with a high degree of connectivity. For example, a forested landscape (matrix) with fewer gaps in forest cover (open patches) will have higher connectivity.

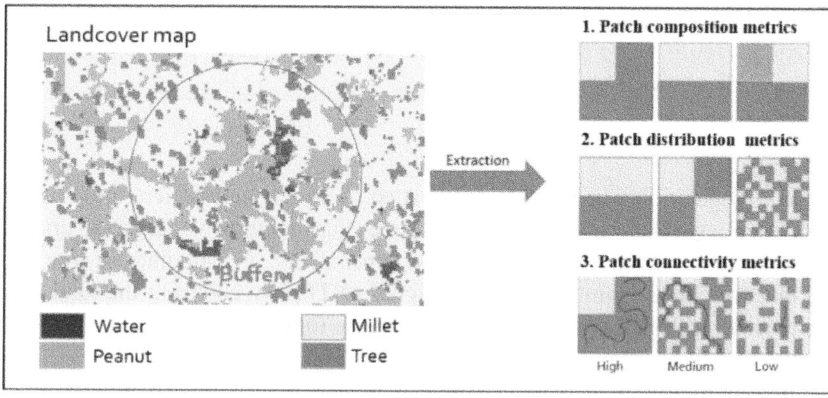

Figure 7.2. *Main landscape indices used to measure the landscape heterogeneity. For a color version of the figure, see www.iste.co.uk/baghdadi/qgis4.zip*

The choice of these indices is generally based on knowledge of the bio-ecology of the targeted species and on the accuracy of available land-use data. But failing that, an exploratory approach can be adopted. It will consist of calculating a large number of landscape variables, which will then be tested statistically. In our case study, we know that the MHM stays in the same millet field during its entire life cycle, and we assume that the surrounding trees shelter the natural enemies. Therefore, we made the choice to select the tree density index and the millet crop density index as pertinent for MHM and the ecology of their natural enemies.

7.2.1.1. *Creation of a hexagonal grid*

Prior to calculation of the landscape variables, we are required to define the size of the landscape unit in which the calculation will be made. In the field of animal biology, the functional area of the target species or its maximum displacement capacity is usually used to define the landscape unit surface area. When this information is lacking, landscape variables are usually calculated at different spatial scales around the sampling sites. In our case study, we have seen that the MHM does not move but its natural enemies can fly a distance of at least 1 km. Therefore, we made the choice to create a grid composed of hexagons, landscape units, with a radius of about 1 km. To do this, we used the "Create Grid" tool of QGIS to generate a hexagonal grid composed of cells of 100 ha. Then the grid was used to cut the land use map into 491 landscape units, which allowed us to calculate the tree density index and the millet crop density index for each cell.

QGIS functionality to create a grid:
=> *Search Create a grid under QGIS Geoalgorithms > Vector creation tools in the Processing Toolbox.*

7.2.1.2. *The tree density index calculation*

Natural vegetation can provide food, such as nectar and pollen, and shelter for natural enemies [OTI 11]. We hypothesize that natural vegetation, and more specifically trees, will enhance MHM biocontrol. Thus, we calculated a tree density index, that we called Number of Tree Patches (NTP), by counting the total number of tree patches (TP) in each hexagonal cell (Hi) as follows:

$$NTP_{Hi} = TP_{Hi} \qquad [7.1]$$

QGIS functionality to calculate the density of tree patches:
=> *Go to Vector>Data Management Tools>Join attributes by location, then Select the "SUM" operator to obtain the number a trees in each hexagonal grid.*

Once calculated for each hexagon of the grid, a tree density map was generated covering the entire study area.

7.2.1.3. *The millet crop density index calculation*

The MHM is a univoltine (one generation per year) noctuid species that undergoes pupal[1] diapause in the soil during the postharvest dry season from October to August. The majority of larvae pupate 10–15 cm deep in heavy soil (clay/loamy) and 15–25 in sandy soil [GAH 86, VER 78]. Therefore, we assume that the MHM population will be more abundant in areas dominated by millet crop and that biocontrol will be less efficient. Conversely, in a more complex landscape composed of other crops, such as groundnut or sorghum and seminatural tree vegetation, the competition of insect species will be promoted and MHM abundance will decrease. Therefore, a millet crop density index, which we called Number of Millet Patches (NMP), representing the relative proportion of millet patches (MP) was calculated for each grid cell (Hi) of the land cover map as follows:

$$NMP_{Hi} = MP_{Hi} \qquad [7.2]$$

QGIS functionality to calculate the density of millet patches:
=> *Go to Vector>Data Management Tools>Join attributes by location, then Select the "SUM" operator to obtain the number a millet patches in each hexagonal grid*

1 The pupa (plural pupae) is the third stage in the life cycle of insects undergoing complete metamorphosis.

Once calculated for each hexagon of the grid, a millet crop density map was generated covering the entire study area.

7.2.1.4. *Classification of landscape metrics data*

The objective of this step is to produce a typology of the millet and tree crop densities of the study area. To do this, we used the Jenks natural breaks classification method [DES 09] in QGIS that makes it possible to obtain three homogeneous classes of density. This is a data clustering method designed to determine the best arrangement of values into different classes. Class breaks are identified that best group similar values and that maximize the differences between classes. The features are divided into classes whose boundaries are set where there are relatively big differences in the data values. Figure 7.3 shows the result of the distribution of the two landscape metrics data (Millet and tree density index) split into three classes of density: low, medium and high density.

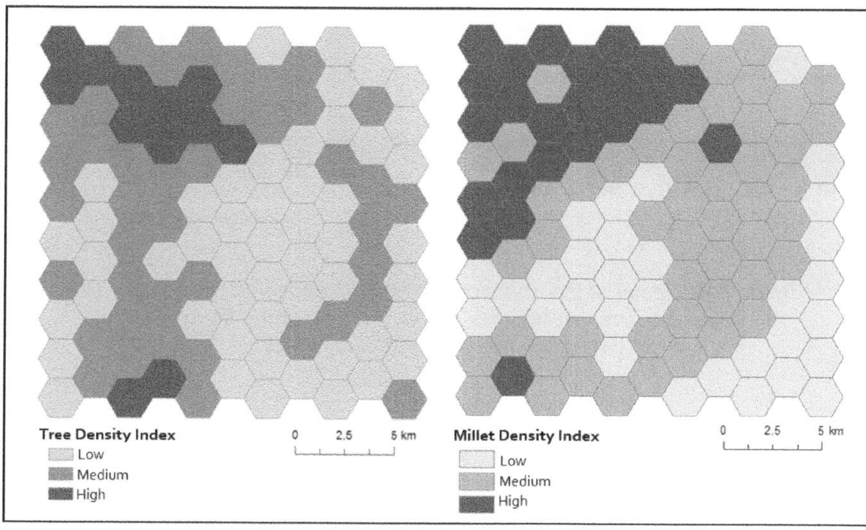

Figure 7.3. *Classification of millet crop density and tree density variables calculated in a cell of 1 km radius. For a color version of the figure, see www.iste.co.uk/baghdadi/qgis4.zip*

7.2.2. *Step 2: sampling plan production*

7.2.2.1. *Combination of landscape variables*

The combination of landscape variables requires prior coding of three density classes of the two maps of landscape variables performed before. A value of "1" was attributed to the low density classes, “2” for the medium density classes , and “3” for the high density classes. Then, all three density classes of both landscape indices were combined in pairs (3 × 3 density classes) making it possible to produce a grid map composed of nine landscape patterns (Figure 7.4). In our case, we obtained only eight landscape patterns, as one combination, (vgt3-m1) between the high density of tree (vgt3) and the low density of millet (m1), did not exist in the study area.

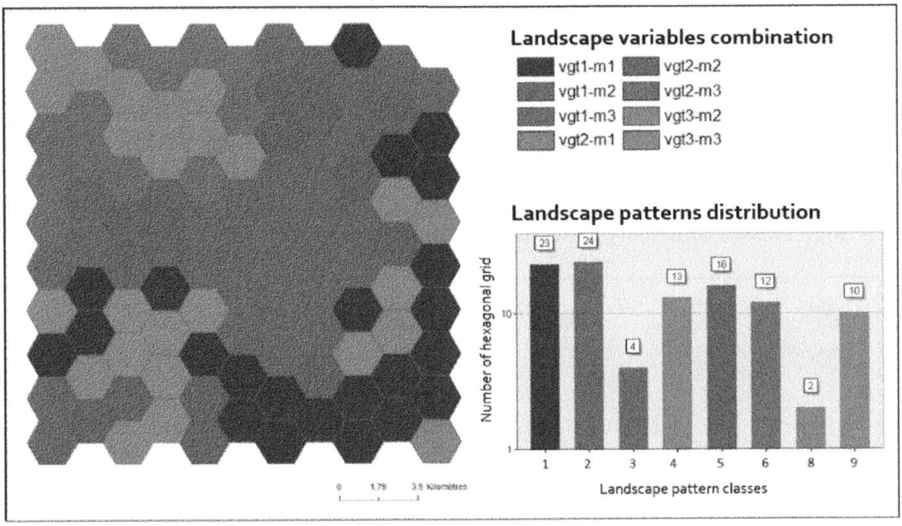

Figure 7.4. *Result of the combination of landscape variables (vgt = tree vegetation, m = millet). For a color version of the figure, see www.iste.co.uk/baghdadi/qgis4.zip*

7.2.2.2. Selection of sampling sites

Then, 35 sampling sites respecting five replications in each landscape pattern (8) were randomly selected from the resulting cluster (Figure 7.5). For this we used the random polygon selection tools in QGIS that allowed us to specify the number of replications for each landscape pattern class. An intersite distance of 2 km between sampling sites was also fixed to avoid any risk of spatial correlation.

QGIS functionality to randomly select sites within landscape patterns:
=> *Click on Vector->Research Tool->Random selection within subsets....*

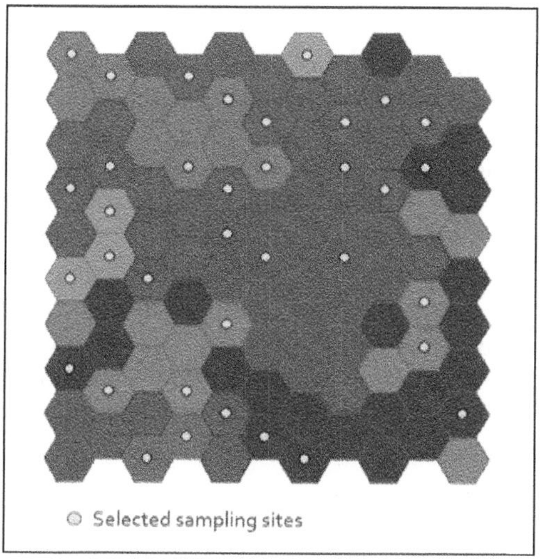

Figure 7.5. *Selected sampling sites in yellow color. For a color version of the figure, see www.iste.co.uk/baghdadi/qgis4.zip*

7.2.3. *Step 3: exportation of selected sampling sites to a GPS*

The selected sampling sites were then exported to a GPS receiver with the objective of validating 35 selected landscape patterns on the field, in terms of tree and millet crop density.

7.3. Practical application

This section presents the methodology used to perform the spatial sampling design for the study of the natural regulation of the MHM in the Bambey area, which covers approximately 400 km^2 in north-east Senegal.

7.3.1. *Software and data*

7.3.1.1. *Required software*

The implementation of the sampling plan based on landscape criteria needs vector layers and the basic functionality of QGIS (version 2.18.3). This application allows anybody to apply the different processing steps without needing to install specific extensions.

7.3.1.2. *Input data*

This practical exercise uses land cover data in vector format, which has been derived from a Pleiades satellite image offering a submetric spatial resolution (<2 m). Validated with ground truth data, the accuracy of the classification was measured using the Kappa (K) coefficient whose value ranges between –1 and 1 [LAN 77]. It is a measure of how well a classification map and the associated reference data agree with each other. This agreement is based on the major diagonal of the error matrix and a chance agreement (row and column values). Strong agreement occurs if the K is greater than 0.80. Moderate agreement occurs when K values fall between 0.40 and 0.80, and poor agreement occurs with K values less than 0.40 [JEN 15]. With a K value of 0.82, the result of the validation is considered to be significant and allows us to obtain a map with eight classes as follows: millet, peanut, market gardening, fallow land, tree vegetation, habitats, roads and bare soils [NDA 14].

These data were obtained within the framework of the TRECS project (Contribution of remote sensing for millet pest control in Senegal), which was financed by the French National Center for Space Studies (CNES). As part of this exercise, we will use a shape file in vector format, named Occsol_Bambey.shp, which has been projected in the Universal Transverse Mercator coordinate system (UTM WGS 84, 28N).

7.3.2. *Step 1: landscape variables calculation*

Step 1 is dedicated to the calculation of landscape variables such as tree vegetation density and millet crop density, according to the two assumptions described previously (sections 7.2.1.1 and 7.2.1.2). To achieve this, several preliminary steps have to be performed: the extraction of target landscape variables (section 7.3.2.1), the estimation of their surface areas (section 7.3.2.2) and the creation of a hexagonal grid composed of cells of about 100 ha (section 7.3.2.3).

7.3.2.1. *Extraction of trees and millet layers*

We will first isolate the "trees" and "millet crops" layers from the land cover map of the study area in vector format.

Process	Practical implementation in QGIS
1. Open the land cover map	In **QGIS**: • Go to Layer and click on Add Vector Layer. • Browse to the `Occsol_Bambey.shp` file and click Open.

<table>
<tr>
<td>2.Inform the coordinate reference system</td>
<td>

- Next you will be prompted to choose the Coordinate Reference System (CRS) => Select EPSG32628, WGS 84, UTM zone 28N as the CRS

</td>
</tr>
<tr>
<td>3. Tree layer extraction</td>
<td>

- Open the Attribute Table of `Occsol_Bambey.shp`. Then, right-click on the layer name and select Open Attribute Table.
- Click on the  icon to perform a query by expression on the `Occsol_Bambey` attribute table.

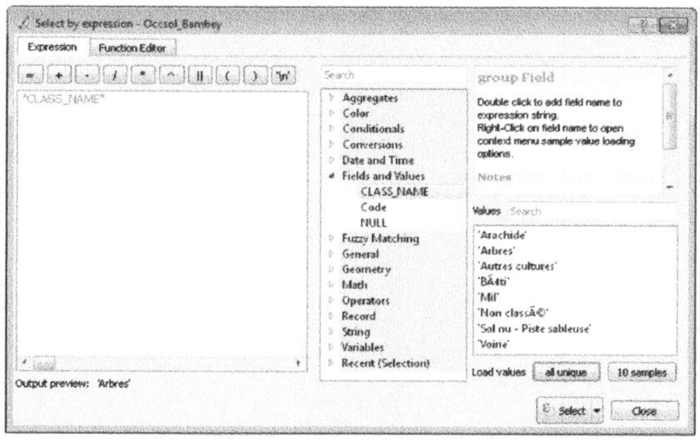

- In the Select by Expression window, expand the **Fields and Values** section and double-click the **Class Name** label. You will notice that it is added in green to the expression section on the bottom left.
- To display the attribute values of the Class Name label, choose the **all unique** option in the Load values section on the bottom left.
- To select the "tree" attribute, click on the icon , and then select the tree class in the list of attributes as follows:

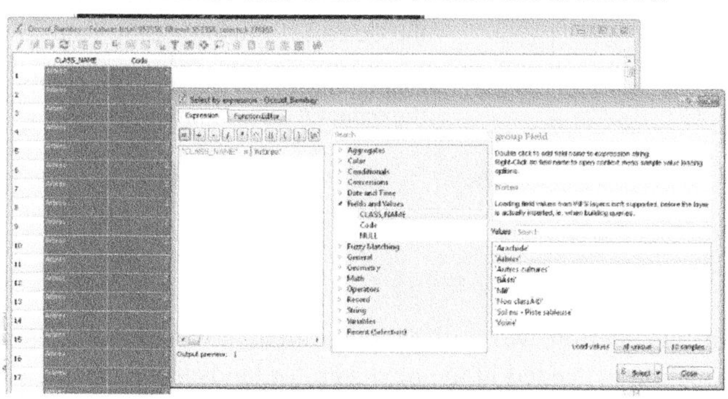

- Click **Select** and **Close** to start the query.

</td>
</tr>
</table>

<table>
<tr>
<td></td>
<td>

- Once the request is performed, double-click on the layer name `Occsol_Bambey.shp`, and select ***Save as...***
- A Save vector layer as window opens. Write "**Tree**" for the file name, check the boxes to "**Save only selected features**" and to "***Add save file to map***" as shown below:

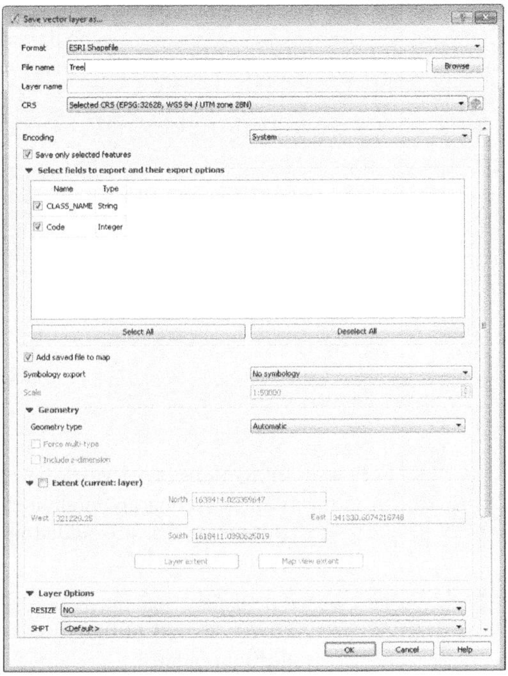

- Click **[OK]** and the Tree shape file will appear in the main QGIS Window.

</td>
</tr>
<tr>
<td>4. Millet layer extraction</td>
<td>

- In the same way, extract the shape file **Millet.shp** from the `Occsol_Bambey.shp` land cover map.
- Then, remove the `Occsol_Bambey.shp` shape file from the main QGIS window.

</td>
</tr>
</table>

Table 7.1. *Millet and tree features extraction*

7.3.2.2. Calculation of areas covered by trees and millet crops

In this section, we will estimate areas covered by trees and millet crops. This information is needed in the next step for landscape variable calculation.

Process	Practical implementation in QGIS
1. Tree area estimation	In **QGIS**: • Open the ***Field calculator*** tool to perform calculations on the basis of existing attribute values or defined functions, e.g. to calculate length or area of geometry features. This tool is directly accessible from the QGIS Standard toolbar by clicking on the icon , or from the Attribute Table of the layer. • Once the ***Field calculator*** window is open: 1) Check the box to **Create a new field** that you will name "Tree_area". 2) Leave the data type in "integer" format. 3) From the menu list at the center of the section, select **Geometry**, and then double-click on "**$ area**" which will then appear in the left window as shown below: 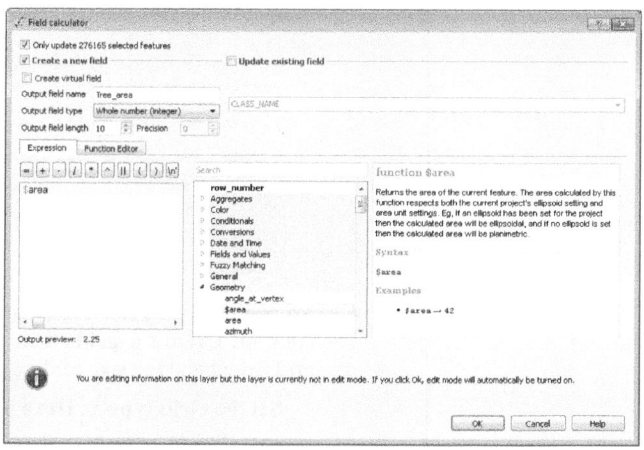 4) Click on the icon **[OK]** to start calculating the areas covered by trees, which will appear in the new field of the attribute table. • On the attribute Table toolbar, turn **Editing Mode** off and then **Save Changes**.
2. Millet area estimation	• In the same way, calculate the surface areas occupied by millet crop and type "Millet_area" for the new field.

Table 7.2. *Millet and tree surface area calculation*

7.3.2.3. *Creating a hexagonal vector grid*

We will build a vector grid composed of hexagonal cells that will be 2 km in diameter. The grid squares that fall outside of the area of interest will be eliminated.

Process	Practical implementation in QGIS
1. Create a grid	• If not, to open the **Processing Toolbox** window, select **Processing** from the standard toolbar, and click **Toolbox.** • To build a grid, search **Create a grid** under **QGIS geoalgorithms** > **Vector creation tools** in the **Processing Toolbox**. 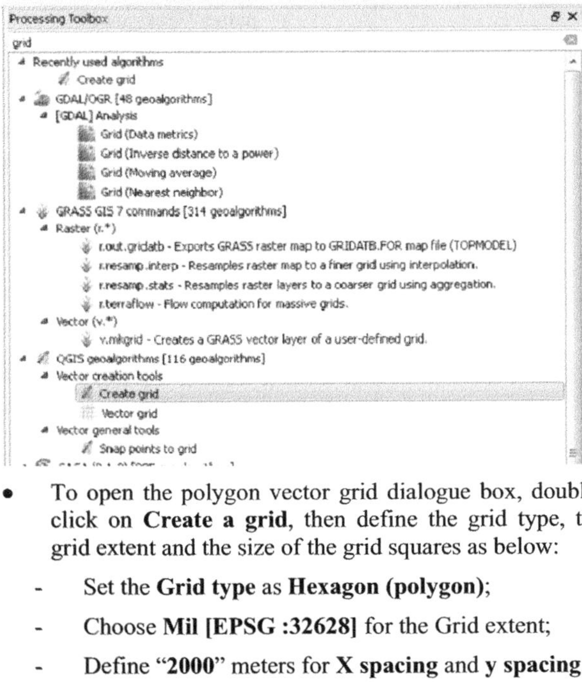 • To open the polygon vector grid dialogue box, double-click on **Create a grid**, then define the grid type, the grid extent and the size of the grid squares as below: - Set the **Grid type** as **Hexagon (polygon)**; - Choose **Mil [EPSG :32628]** for the Grid extent; - Define "**2000**" meters for **X spacing** and **y spacing**; - Select "**WGS84 / UTM 28N**" as the coordinate reference system (CRS); - Type "**Grid_2km**" for the **Grid name**; - Click **RUN** to build the grid. • Then, the grid will appear in the main QGIS window. You can change the position of the grid above the Trees and Millet features by dragging "Grid_2km" using the mouse cursor.

	• Select the `Grid_2 km.shp` layer from the **Layers Panel** window, click **Properties** and then **Style** to change the polygon fill to transparent; thicken the outline to make it more visible. • Select **Simple fill** in the window section on the top left. Then, as shown below, choose the **No brush** option as **Fill Style** and the **red** color for **Outline**. 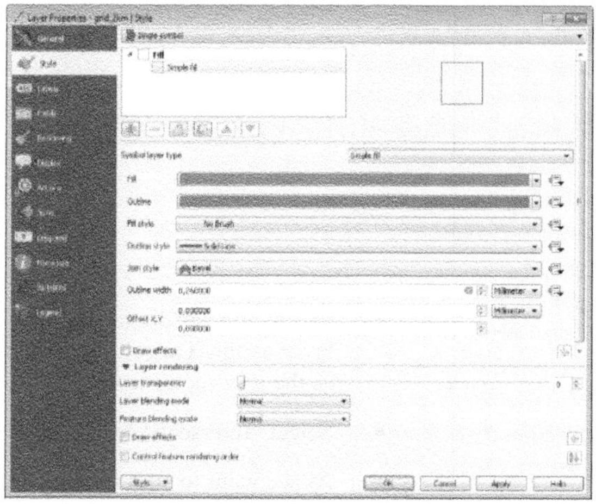 • Click **OK** and **Apply** to display the new grid.
2. Create field area	• Now we will calculate the area of each grid square in a field. Therefore, as described in section 7.3.2.2 , use the ***Field calculator tool*** by selecting the button 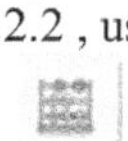 , from the `Grid_2km.shp` attribute table and create a new field named "Grid_area" that you will fill by selecting the ***$area*** operator proposed in the ***Field calculator*** window. • **Save** your edits and **end** the editing session.

Table 7.3. *Generate a hexagonal grid. For a color version of the table, see www.iste.co.uk/baghdadi/qgis4.zip*

7.3.2.4. *Landscape variables calculation per cell*

We will calculate the number of trees and the number of millet patterns in each cell of the grid as density variables. Therefore, we merge data from two different features, "Grid_2km.shp" and the two targeted landscape layers, "Tree.shp" and "Millet.shp".

Process	Practical implementation in QGIS
1. Number of trees per cell calculation	• Go to ***Vector>Data Management Tools>Join attributes by location***: 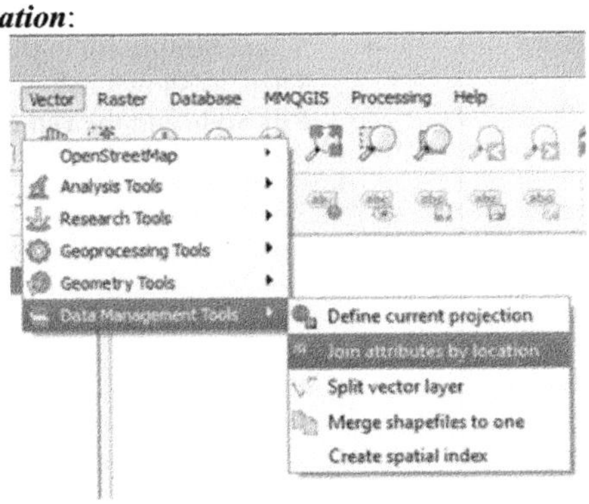 • In the join attribute dialogue box, select the `Grid_2km` layer as the **Target vector layer** and the `Tree.shp` layer as the join vector layer. • Select "Intersects" as Geometric predicate. • As we want to sum the number of trees per cell, select Take summary of intersecting features and choose Sum. • Name the output file as "`Tree_Grid.shp`". • In the Output table select Only keep matching records. 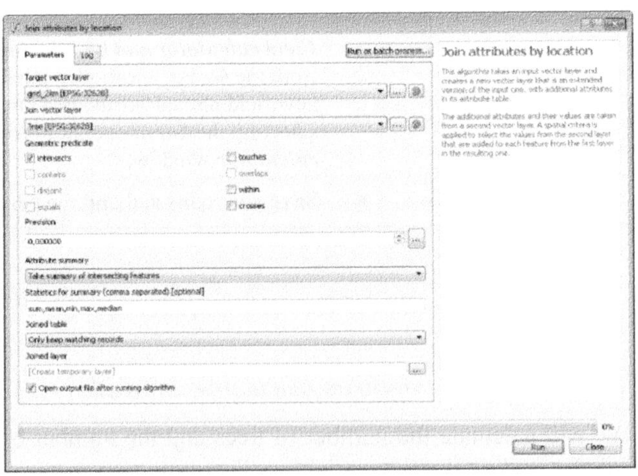 • Click Run and once the process finishes, select Yes when asked if you want to add the layer to Table of Contents (TOC).

2. Tree density classification	• Right-click on the layer name `Tree_Grid.shp` and select Properties. You can also double-click on the layer name in the TOC to bring up the Layer Properties dialog. • In the Layer Properties window, select Style. • Choose a Graduated classification type. • Select the column COUNT and choose the color ramp YlOrR to generate a new color map. • Choose 3 as the number of classes and Natural Breaks (Jenks) as the data classification method. 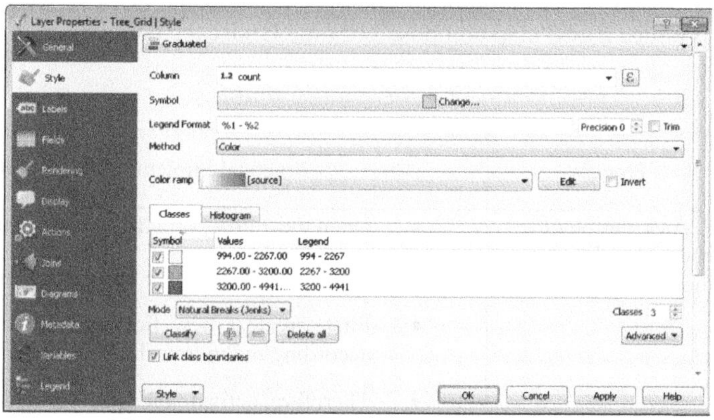 • Click OK and Apply
3. Calculation of the number of millet patterns	• As seen for the tree case before, go to Vector>Data Management Tools>Join attributes by location. • In the dialogue box, select the `Grid_2km` layer as the Target vector layer and the `Millet.shp` layer as the join vector layer. • As we want to sum the number of millet patches per cell, select Take summary of intersecting features and choose Sum. • Name the output file `Millet_Grid.shp`. • In the Output table, select **Only keep matching records**. • Click **Run** to start the process.
4. Classification of millet patterns' density	• Right-click on the layer name `Millet_Grid.shp` and select Properties>Style. • Choose a Graduated classification type. • Select the column **COUNT** and choose the color ramp Greens to generate a new color map. • Choose 3 as the number of classes again and Natural Breaks

	(Jenks) as the data classification method. • Click OK and Apply.

Table 7.4. *Landscape metrics calculation per cell grid. For a color version of the table, see www.iste.co.uk/baghdadi/qgis4.zip*

7.3.3. *Step 2: sampling plan production*

In the second step, all three density classes of both landscape indices' classifications are combined in pairs (3 × 3 density classes), which makes it possible to produce a grid map composed of nine landscape patterns.

7.3.3.1. *Combination of landscape variables*

Process	Practical implementation in QGIS
1. Assignation of a code according to the tree class density	• Open the `Tree_Grid.shp` **attribute table**, then click on the icon to display the ***Field calculator*** tool box. • First, create a new field that we will fill with a code according to the tree class density. • Therefore, activate the **Editing mode** by clicking on the icon, and then click on the icon to create a new field. Type "**Cod_vgt**" as the output field name and choose "**Text (String)**" for the output field type. • Now we are going to perform the query to assign the code "vgt1" to low density of trees, "vgt2" to medium density of trees, and "vgt3" to high tree density. • Therefore, in the ***Field calculator*** window, check "***Update existing file***", and choose the targeted field "**Cod_vgt**" to be filled, and then enter the expression as shown below in the expression section: CASE WHEN "count" >= 994 AND "count" <= 2267 THEN 'vgt1' WHEN "count" > 2267 AND "count" <= 3200 THEN 'vgt2' WHEN "count" > 3200 AND "count" <= 4941 THEN 'vgt3' END

<table>
<tr>
<td></td>
<td>
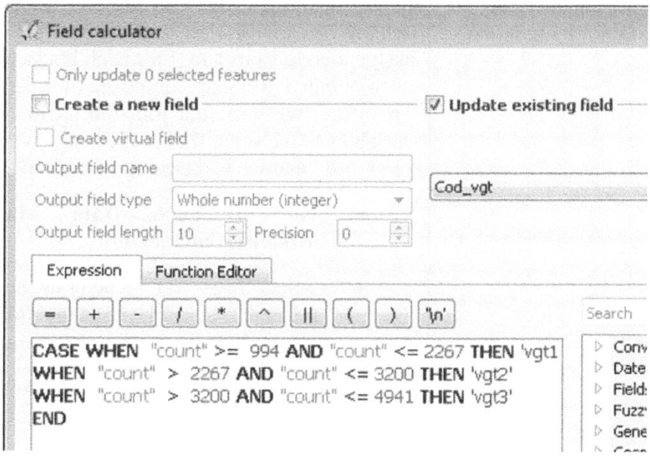

<ul>
<li>Values of the border classes arise from the landscape variables' classification performed before in section 7.3.2.4.</li>
<li>Click OK, and the new field Cod_vgt will automatically fill up.</li>
<li>Once the process is finished, turn off the Editing mode and Save changes.</li>
</ul>
</td>
</tr>
<tr>
<td>2. Assignation of a code according to the millet class density</td>
<td>
<ul>
<li>Repeat the same process for the <code>Millet_Grid.shp</code> layer. Create a new field name "Cod_mil", then assign the code class (m1, m2, m3) according to the millet class density as follows:</li>
</ul>
CASE WHEN "count" >= 61 AND "count" <= 338 THEN 'm1'
WHEN "count" > 338 AND "count" <= 650 THEN 'm2'
WHEN "count" > 650 AND "count" <= 1080 THEN 'm3'
END
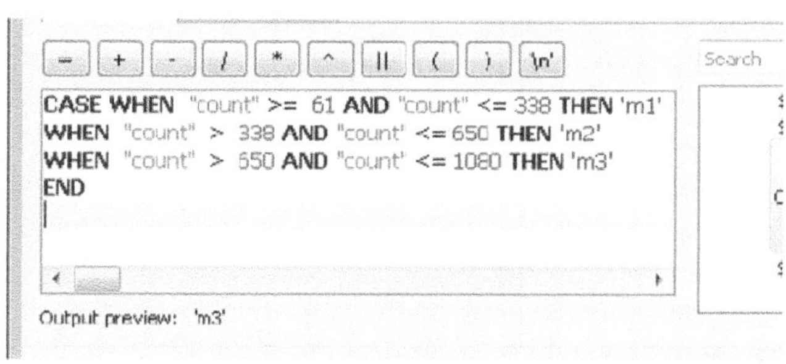

<ul>
<li>Once the process is finished, do not forget to turn off the Editing mode and to Save changes.</li>
</ul>
</td>
</tr>
</table>

<table>
<tr>
<td>3. Join attribute tables by location</td>
<td>In the next step, we will combine each class of density (3 × 3 density classes) of both landscape variables in pairs. Then, we will assign a code from 1 to 9 for each landscape combination, allowing us to obtain a grid map composed of nine landscape patterns.
Therefore, we will first join the attributes of both the landscape variables in the same file, in order to have the fields "cod_vgt" and "cod_mil" that we just created in the same attribute table.

• Go to Vector>Data Management Tools>Join attributes by location:
• Select <code>Tree_Grid</code> layer as the Target vector layer and <code>Millet_Grid</code> layer as the Join vector layer.
• Check Equals as Geometric predicate.
• Select "Take attributes of the first located feature" as attribute summary and the joining option "Only keep matching records".
• Do not choose a statistical operator.
• Name the output file <code>Combi.shp</code>.
• Click Run to start the process.
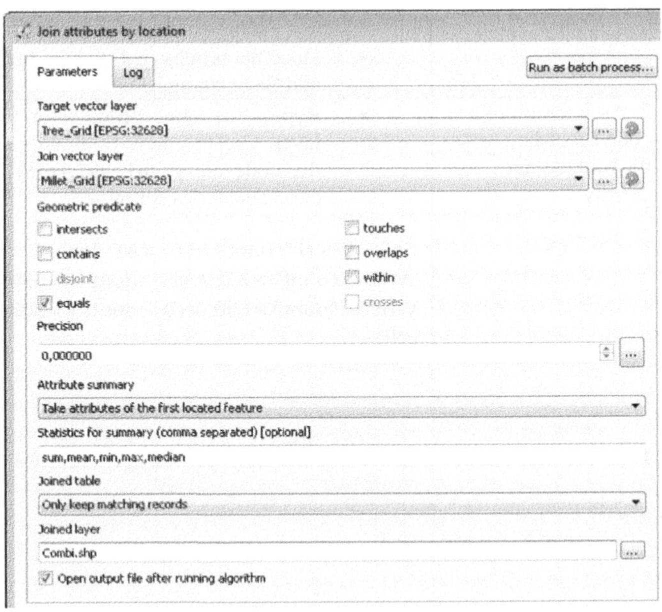

</td>
</tr>
</table>

<table>
<tr>
<td>4. Assignation of a code according to the landscape pattern</td>
<td>
Now, we will create a new field named “code_combi” in the attribute table of the <code>Combi.shp</code> layer.

• As described before, use the Field calculator tool and in the window expression enter the following code lines:

CASE WHEN "cod_vgt" = 'vgt1' AND "cod_mil" = 'm1' THEN 1
WHEN "cod_vgt" = 'vgt1' AND "cod_mil" = 'm2' THEN 2
WHEN "cod_vgt" = 'vgt1' AND "cod_mil" = 'm3' THEN 3
WHEN "cod_vgt" = 'vgt2' AND "cod_mil" = 'm1' THEN 4
WHEN "cod_vgt" = 'vgt2' AND "cod_mil" = 'm2' THEN 5
WHEN "cod_vgt" = 'vgt2' AND "cod_mil" = 'm3' THEN 6
WHEN "cod_vgt" = 'vgt3' AND "cod_mil" = 'm1' THEN 7
WHEN "cod_vgt" = 'vgt3' AND "cod_mil" = 'm2' THEN 8
WHEN "cod_vgt" = 'vgt3' AND "cod_mil" = 'm3' THEN 9
END
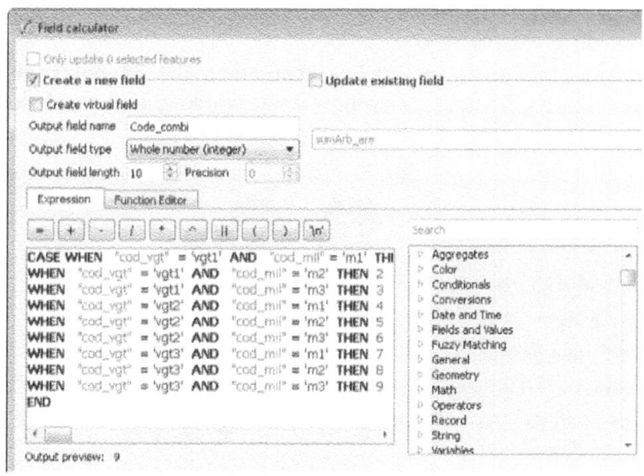

• Once the process is finished, turn off the Editing mode and Save changes.
</td>
</tr>
</table>

Table 7.5. *Landscape variables combination. For a color version of the table, see www.iste.co.uk/baghdadi/qgis4.zip*

7.3.3.2. *Random selection of sampling points*

We are going to randomly choose the sampling points from the result, which has eight landscape pattern classes, as explained in section 7.2.2.1. Then, respecting the constraint of five replications per landscape pattern, we select the 35 sampling sites using the random polygon selection tools in QGIS that allows us to specify the number of replications (5) for each landscape pattern class (8). An intersite distance of 2 km between sampling sites has also been fixed to avoid any risk of spatial correlation.

Process	Practical implementation in QGIS
1. Creation of random polygons	• Click on **Vector->Research Tool->Random selection within subsets.** 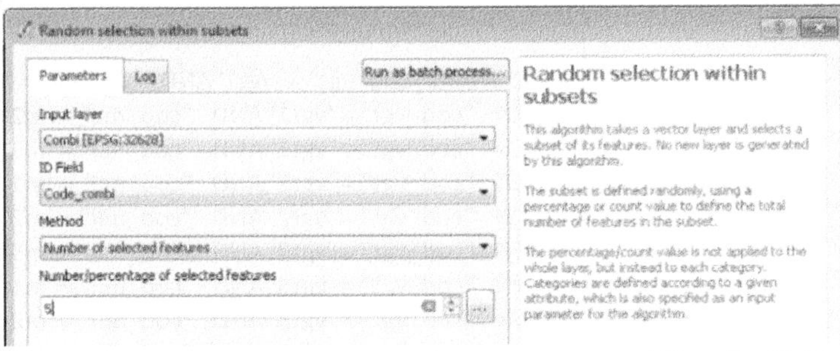 • Select Combi.shp as input layer, "**Code_combi**" as ID_Field, "**Number of selected features**" as the random method, and type **5** for the number of selected features. • Click **Run**, then the **35 polygons** selected will appear in yellow in the window, as shown below: 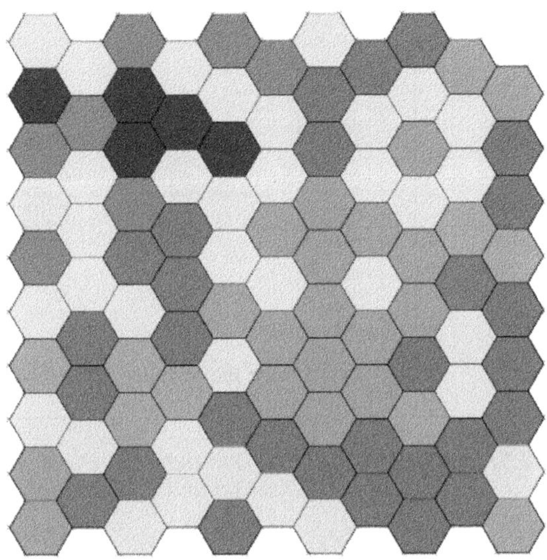
2. Extraction and recording of the polygon selection	• To save the sites selection, right-click on the name of the `Combi.shp` layer, choose **Save as**... and name the polygon selection "`Sampling_poly.shp`". Then, check the option **"Save only selected features" in the encoding section and click OK**.
3. Computation of the polygon centroids	• Open the centroids attribute table by clicking on **Vector<Geometry tools< Polygon centroids**.

<table>
<tr>
<td></td>
<td>
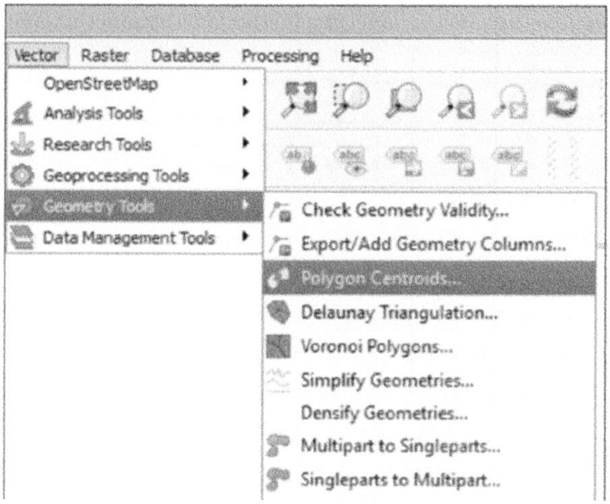

<ul>
<li>Select <code>Sampling_poly.shp</code> as input layer and name the output file <code>Sampling point.shp</code>.</li>
<li>Click RUN to generate the Polygon centroids which will appear in yellow in the QGIS window, as shown below:</li>
</ul>
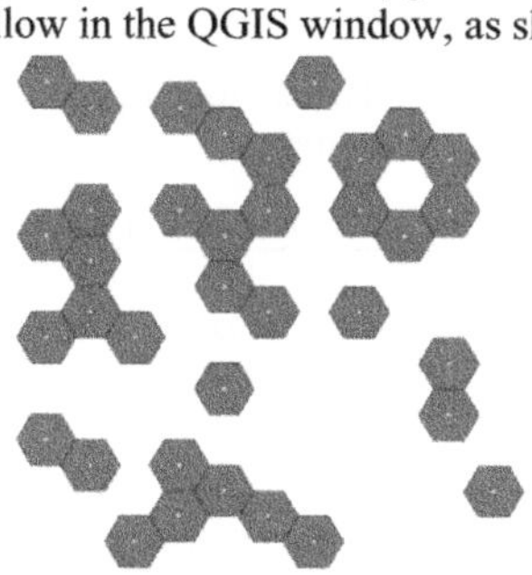
</td>
</tr>
<tr>
<td>4. Addition of an X and Y coordinate to point data</td>
<td>
You will create two new fields, X and Y, where coordinates of each point will be recorded. Remember longitude is always X and latitude is always Y.
<ul>
<li>Open your attribute table by right clicking on the properties of the <code>Sampling_points.shp feature</code>.</li>
<li>Open the Field Calculator.</li>
<li>You will want to use the $x and $y Functions under geometry.</li>
<li>In the field calculator, create a new field called Xcoord.</li>
</ul>
</td>
</tr>
</table>

	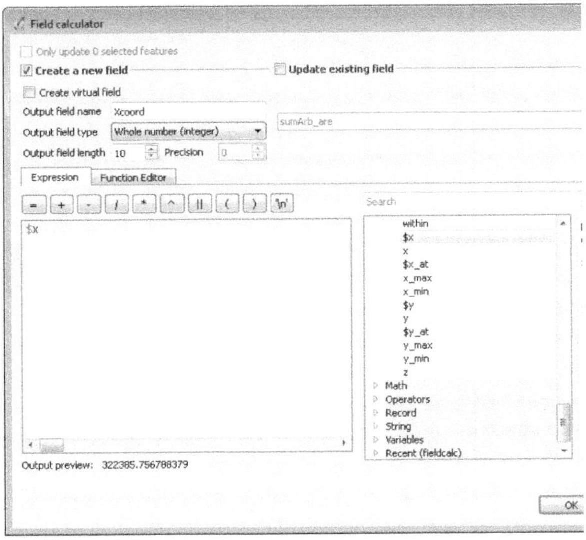 • Double-click on **$x** function and then click **OK** to perform the process. • Repeat to generate a **Y** coordinate. For this you will use the **$y** function. • Click **OK** and **Save** the adding information.

Table 7.6. *Random selection of sampling points. For a color version of the table, see www.iste.co.uk/baghdadi/qgis4.zip*

7.3.4. *Step 3: integrating sampling points into a GPS device*

The selected sampling points are then exported to a GPS device with the objective of validating the 35 selected landscape patterns on the field, in terms of tree and millet crop density.

7.3.4.1. *Converting shape file points into GPX format*

Process	Practical implementation in QGIS
1. Conversion of shape file points into gpx format	• Browse and double-click on "**Convert Format**" located in the **Processing Toolbox**.

	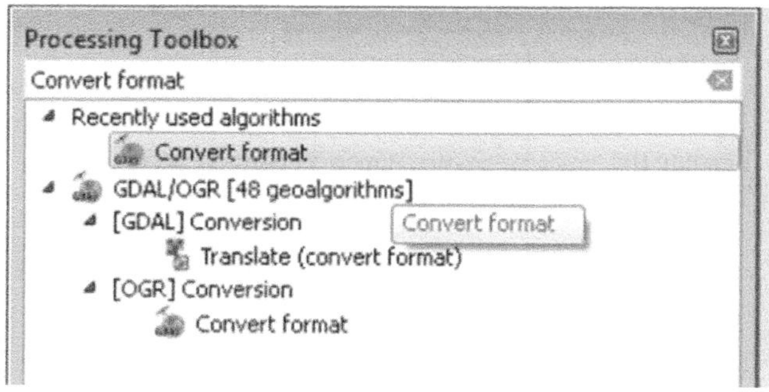 • In the displaying window, select `Sampling_points.shp` as input layer. • Select **GPX** as destination format. • Then Save and Name the new file as `GPSpoints_validation.gpx`. 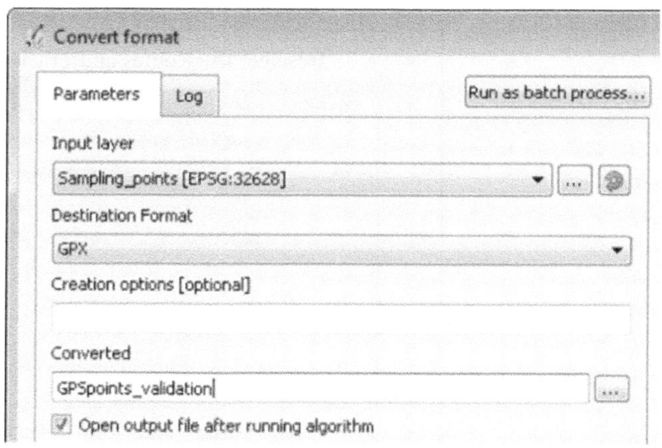 • If the conversion tool does not work, you can use also **GPS Trackmaker** downloadable software, available at this address: http://www.trackmaker.com/main/index.php/en-us/menusoftwares-en.

Table 7.7. *Converting shape file point data into GPX format*

7.3.4.2. *Uploading points onto a GPS device*

<table>
<tr><th>Process</th><th>Practical implementation in QGIS</th></tr>
<tr><td>1. Loading the GPX file</td><td>

- Open **Vector/GPS/GPS Tools.**

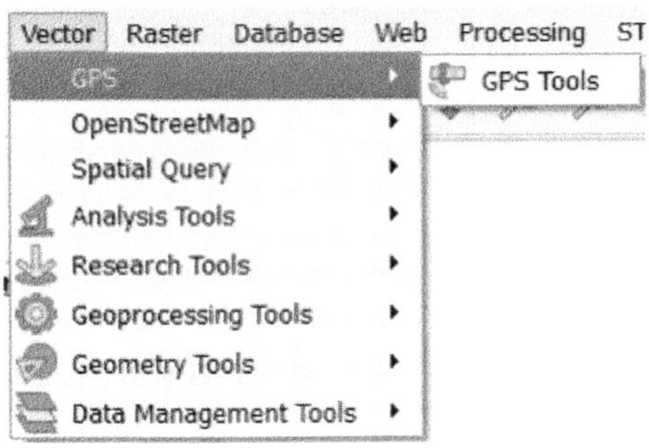

- **Click** on **Load GPX file** and Select `Points_validation.gpx` as input file.
- **Deselect the** Routes and Tracks feature.

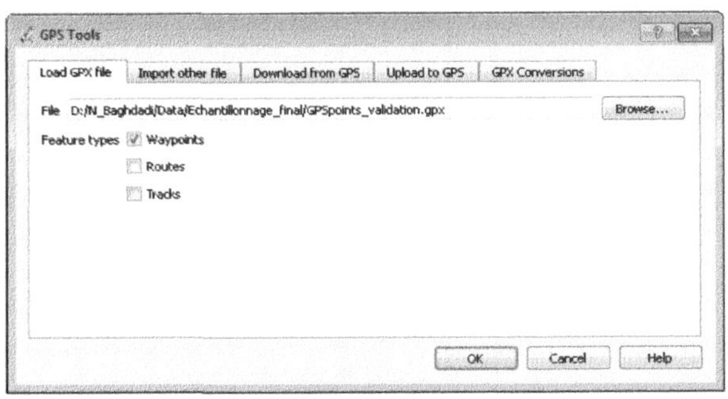

- Click **OK** and the `Points_validation.gpx` feature will be displayed in the layer panels.
- Connect your GPS device to the computer.

</td></tr>
<tr><td>2. Exportation of points to a GPS</td><td>

- Now, you are able to export the GPX file to the GPS device using **Upload to GPS** from the **GPS Tool box** as below:

</td></tr>
</table>

	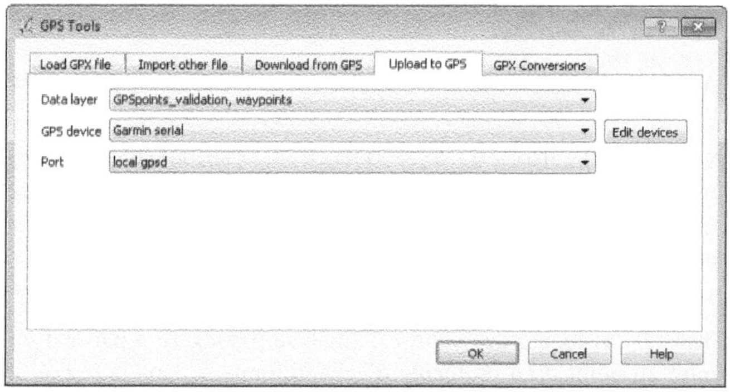 • The `Points_validation.gpx` feature will appear automatically as a data layer. Then, select your GPS device type and the port (or usb) that it is connected to. • Click **OK** to start the exportation. Once your sampling points are uploaded into the GPS device, you can use the **Go to** function which allows you to easily navigate each sampling point for validation.

Table 7.8. *Upload feature points into a GPS device*

7.3.5. *Limits of the method*

The sampling method presented here is adapted when the number of landscape variables is small (two or three maximum) and can easily be performed in GIS software such as the free software QGIS. It is also advisable to limit the number of variables in order to avoid interpretation issues due to combinations of landscape variables, and the need to increase the number of observations on the field. When a large number of environmental variables are to be taken into account, other methods in the statistical domain are more suitable. Roux *et al.* [ROU 13] recently published an article proposing a methodology to perform reduced sampling plans (small plots) in heterogeneous contexts. This involves performing multiple component analysis based on a successive selection of variables (48 attributes in the study) and a variety of sources (environmental, social, economic) to determine a low number of samples, and the selection of the most heterogeneous sites.

7.4. Bibliography

[AJA 80] AJAYI O., "Insect Pests of Millet in Nigeria", *Samaru Miscellaneous Paper*, no. 97, p. 10, 1980.

[BIA 06] BIA F., BOOI J., TSCHARNTKE T., "Sustainable Pest Regulation in Agricultural Landscapes: A Review on Landscape Composition, Biodiversity and Natural Pest Control", *Proceedings of the Royal Society of London B: Biological Sciences*, vol. 273, no. 1595, pp. 1715–1727, 2006.

[BIL 13] BILLAND C., "Systèmes agroforestiers en zone sèche et régulation naturelle des insectes ravageurs des cultures: Analyse paysagère à partir d'une image satellite pour la mise au point d'un protocole d'échantillonnage dans la région de Dangalma au Sénégal, Mémoire d'ingénieur: Gestion des milieux naturels", *AgroParisTech*, p. 143, 2013.

[DES 09] DE SMITH M.J., GOODCHILD M.F., LONGLEY P.A., *Geospatial Analysis*, Matador, Leicester, UK, 2009.

[GAH 86] GAHUKAR R.T., GUEVREMONT H., BHATNAGAR V.S. *et al.*, "A Review of the Pest Status of the Millet Spike Worm, Raghuva Albipunctella De Joannis (Noctuidae: Lepidoptera) and Its Management in the Sahel", *International Journal of Tropical Insect Science*, vol. 7, no. 4, pp. 457–463, 1986.

[GUE 82] GUEVREMONT H., Recherches Sur L'entomofaune Du Mil, Rapport annuel de Recherches, CNRA, 1982.

[JEN 15] JENSEN J.R., *Introductory Digital Image Processing: A Remote Sensing Perspective*, 3rd ed, Prentice Hall, New Jersey, 2015.

[KRA 95] KRALL S., YOUM O., KOGO S., "Panicle Insect Pest Damage and Yield Loss in Pearl Millet", *Proceedings of an International Consultative Workshop on Panicle Insect Pest of Sorghum and Millet*, ICRISAT Sahelian Centre, Niamey, Niger, 1995.

[LAN 77] LANDIS J., KOCH R.E.T., GARY G., "The measurement of observer agreement for categorical data", *Biometrics*, vol. 34, no. 1, pp. 159–174, 1977.

[LAN 00] LANDIS D.A., WRATTEN S.D., GURR G.M., "Habitat management to conserve natural enemies of arthropod pests in agriculture", *Annual Review of Entomology*, vol. 45, no. 1, pp. 175–201, 2000.

[NDO 79] NDOYE M., "L'entomofaune Nuisible Du Mil a Chandelle (Pennisetum Typhoides) Au Sénégal", *Compte rendus des travaux : Congrès sur la Lutte Contre les Insectes en Milieu Tropical*, 13–16 March, 1979.

[OTI 11] OTIENO M., WOODCOCK B.A., WILBY A. *et al.*, "Local Management and Landscape Drivers of Pollination and Biological Control Services in a Kenyan Agro-Ecosystem", *Biological Conservation*, vol. 144, no. 10, pp. 2424–2431, 2011.

[ROU 13] ROUX E., GABORIT P., ROMAÑA C.A. *et al.*, "Objective Sampling Design in a Highly Heterogeneous Landscape – Characterizing Environmental Determinants of Malaria Vector Distribution in French Guiana, in the Amazonian Region", *BMC Ecology*, vol. 13, no. 1, pp. 45–58, 2013.

[WIL 11] WILLIAM P., TAPSOBA H., BAOUA I.B. *et al.*, "On-Farm Biological Control of the Pearl Millet Head Miner: Realization of 35 Years of Unsteady Progress in Mali, Burkina Faso and Niger", *International Journal of Agricultural Sustainability*, vol. 9, no. 1, pp. 186–193, 2011.

[YOU 98] YOUM O., OWUSU E.O., "Assessment of Yield Loss Due to the Millet Head Miner, Heliocheilus Albipunctella (Lepidoptera: Noctuidae) Using a Damage Rating Scale and Regression Analysis in Niger", *International Journal of Pest Management*, vol. 44, no. 2, pp. 119–121, 1998.

8

Modeling Erosion Risk Using the RUSLE Equation

8.1. Definition and context

Erosion processes include all the drafting and transportation mechanisms of sediments at the Earth's surface. They are either physical or chemical. Erosion agents can be meteorological, hydric, gravity-related or human. Knowledge, estimation and management of erosion are key factors in the management of economic (agronomic quality of soils, economic vulnerability), environmental (river discharge, biodiversity loss) and social (water drinking supply, risks to lives and goods) issues.

There are many erosion models, which may or may not include the spatial dimension of processes, to qualify and quantify erosion. One of the most commonly used spatial models is the empirical model relying on the Universal Soil Loss Equation [WIS 78]. This model has the advantages:

– of exploiting accessible geographical parameters (climate, land cover, topography);

– of being operated using GIS software;

– of being adaptable to the degree of knowledge of the area studied: it can thus be exploited quantitatively or qualitatively depending on the validation given to each parameter;

– of being able to provide quantitative estimates of soil loss.

Chapter written by Rémi ANDREOLI.

The model had been revised since its first release. The Revised Universal Soil Loss Equation (RUSLE) relies on the same equation but the methods to obtain the different parameters have been improved.

8.2. RUSLE model

The RUSLE model, initially used to quantify soil losses over agricultural parcels, provides estimates of sediment displacement within each cell (or pixel) of the model due to hydric processes (rainfall and surface runoff). The soil losses are expressed in [Mass]·[Surface Unit]$^{-1}$·[Time period]$^{-1}$ such as ton/ha/year.

The RUSLE equation is the following:

$$A = R \times K \times L \times S \times C \times P \quad [8.1]$$

– **A** is the soil loss expressed in the same unit as the K factor [Mass]·[Surface unit]$^{-1}$ and for the period selected for the R period. Theoretically, the soil loss varies between 0 and infinity; in practice, splash and sheet erosion correspond to A values between 0 and 5 ton/ha/year and erosion inside lavakas[1] or gully erosion can easily exceed 1,000 ton/ha/year;

– **R** is the rainfall aggressiveness factor. R varies between 0 and infinity; under tropical climates, R factor mean values reach 1500-–1900 MJ·mm·(ha·h·year)$^{-1}$, and can exceed 3,000 MJ·mm·(ha·h·year)$^{-1}$ [DUM 10a];

– **K** factor characterizes soil erodibility. This dimensionless factor is experimentally obtained and provides soil loss capacity in ton per surface per R unit for each soil. Usually, R varies from 0 to 0.7 (t·acre·year)/(ha·MJ·mm) ;

– **L** is the slope length expressed in length unit;

– **S** is the slope gradient. For convenience, S factor is estimated jointly with L factor in a unique LS factor. LS is dimensionless and varies between 0 and infinity;

– **C** is the vegetation cover factor. It is also a dimensionless factor that varies between 0 and 1;

– **P** factor qualifies agricultural and soil protection practices. P is dimensionless and varies between 0 and 1.

1 Lavaka is a malagasy word meaning "hole" and used worldwide in geomorphology to characterize deep and egg-shaped systems of gullies opened in lateritic soils.

The six parameters of the RUSLE model are directly derived or estimated from the following data:

– rainfalls – R factor;

– digital elevation model (DEM) – LS factor;

– land use and land cover databases – K, C and P factors.

The implementation of the RUSLE model consists of three main steps (Figure 8.1):

– the preprocessing of the data consists of data formatting. This step includes reprojection of data, clipping of the data according to the area of interest, some attribute correction and rasterizing;

– indicators' estimates are then realized according to theoretical formulae or according to attribute assignment rules;

– the soil loss estimate is calculated using formula [8.1].

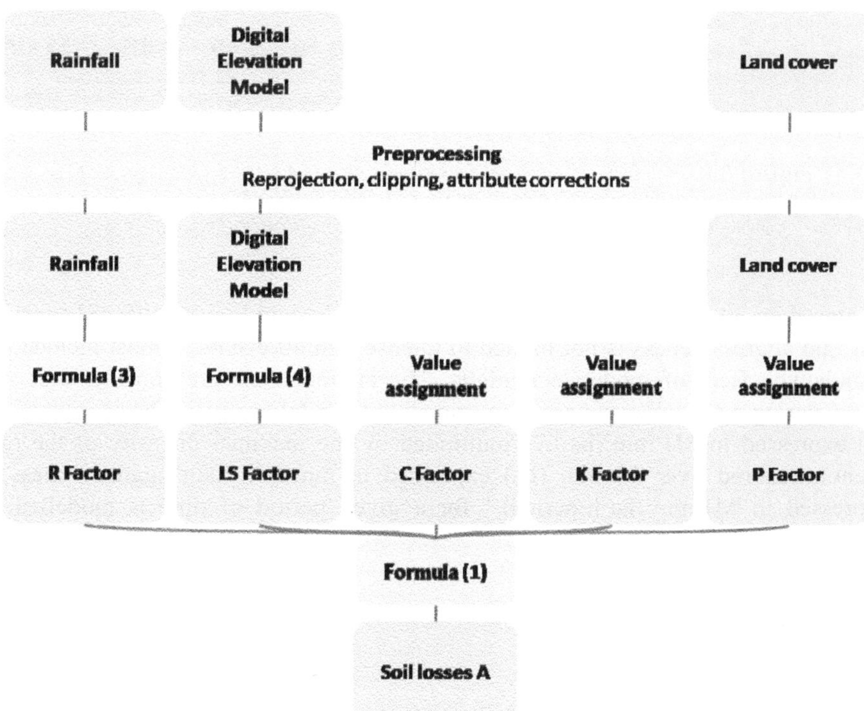

Figure 8.1. *RUSLE modeling processing steps*

8.2.1. *Climatic factor: rainfall aggressiveness R*

Rainfall aggressiveness (Figure 8.2) is a factor that determines the capacity of rainfall events to displace soil. The R factor therefore takes into account total rainfall and the occurrence of rainfall peaks capable of causing soil displacement.

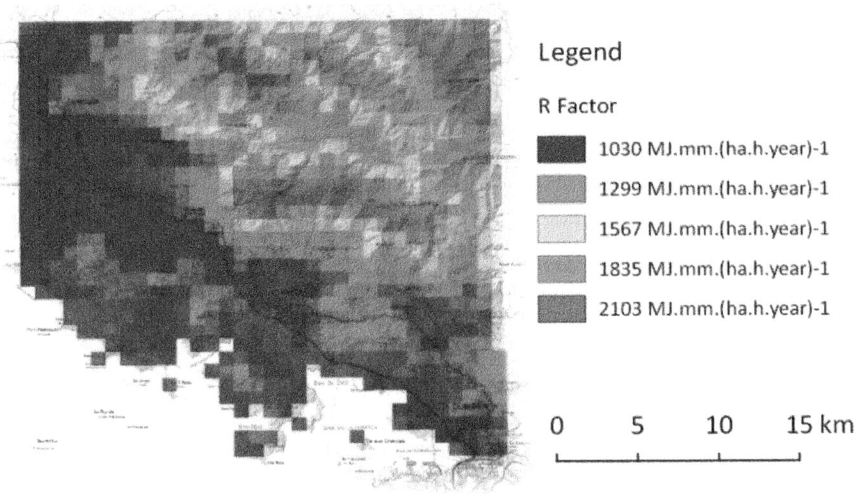

Figure 8.2. *R factor, rainfall aggressiveness in MJ·mm·(ha·h·year)$^{-1}$ over the area of Dumbéa—New Caledonia (166.3478°W, 22.0652°S). For a color version of the figure, see www.iste.co.uk/baghdadi/qgis4.zip*

Based on 30 years of rainfall records, Wischmeier and Smith [WIS 78] consider that rain aggressiveness is not limited to intense rainfall events. It must include the cumulative effect of moderate rainfalls. The authors therefore consider that soil losses during a rainfall event are proportional to the total energy of the precipitations (E) expressed in MJ·mm·(ha·h)$^{-1}$ multiplied by the maximal intensity of the rain event measured over 30 min (I_{30}) expressed in mm·h^{-1}. Rain aggressiveness R expressed in MJ·mm·(ha·h·period)$^{-1}$ for a given period of time is modelled as follows [REN 94]:

$$R = \frac{\sum_{i=1}^{j}(EI_{30})_i}{N} \quad [8.2]$$

with:

– $(EI_{30})_i = EI_{30}$ for the rain event i;

– j is the number of rain events;

–N is the period considered (a month, a year, several decades, etc.).

Nevertheless, the use of this formula to assess the spatial repartition of erosion requires accurate rainfall measurements covering the entire area of interest, which is often difficult to obtain. Furthermore, the authors propose several approximation methods of the R factor according to the parameters available. Roose [ROO 77] developed an approximation based on the following equation:

$$R = k \times P \times 0.5 \tag{8.3}$$

with:

– P is mean annual rainfalls in $MJ \cdot mm \cdot (ha \cdot h \cdot year)^{-1}$;

– k is an adjustment coefficient. We use a k value of 1.73 in this chapter, proposed by [DUM 10a].

QGIS functionality for rainfall aggressiveness estimates from mean annual rainfall:
- Raster calculator... : *Raster > Raster calculator...*

8.2.2. *Topographic factor: slope length and gradient*

The slope length L and the slope gradient S derived from topography both affect soil sensitivity to erosion processes due to rainfalls. Although they were initially assessed separately, the use of a unique LS factor is more convenient [WIS 78] (Figure 8.4).

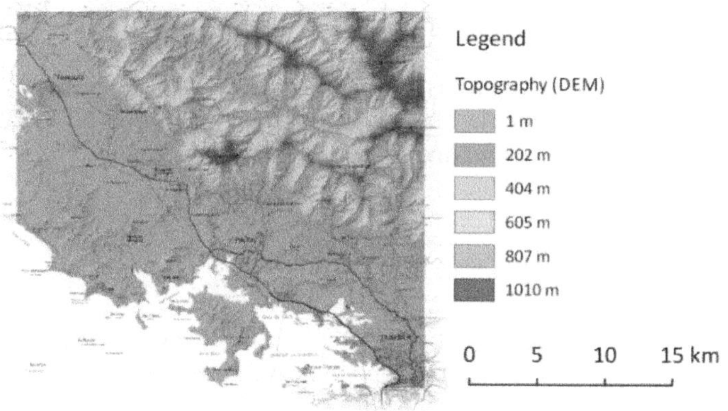

Figure 8.3. *Digital Elevation Model (DEM) over the area of Dumbéa — New Caledonia (166.3478°W, 22.0652°S). For a color version of the figure, see www.iste.co.uk/baghdadi/qgis4.zip*

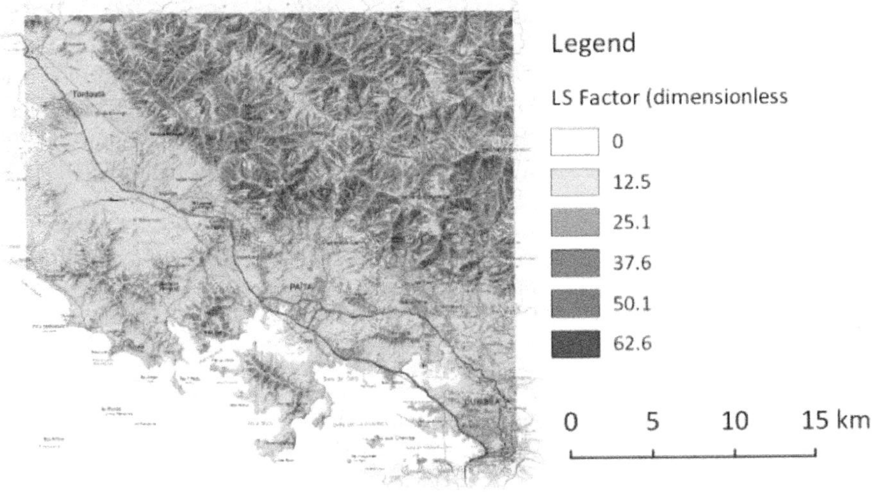

Figure 8.4. *LS factor (L is the slope length and S is the slope gradient) over the area Dumbéa – New Caledonia (166.3478°W, 22.0652°S). For a color version of the figure, see www.iste.co.uk/baghdadi/qgis4.zip*

Several algorithms were developed to estimate the LS factor [DES 96, MOO 89, MOO 91, PAN 91, TAR 05, WIS 78]. The first formula takes into account the slope length upstream of each point of the DEM [WIS 78]. Next, formulas more commonly found in GIS software replace the slope length with the contributive area upstream of each point of the model. The differences between equations therefore derive from the application of selected coefficients to better account for the relief of the study area (plain, plateau, hills, mountains).

In areas where the topography is complex (mix of plateaus, terraced ledges and mountains) as in New Caledonia, the equation proposed by [MOO 89] better takes into account the convergence and divergence of the flows. Thus, *LS factor is defined by:*

$$LS = \left[\frac{A_S}{22.13}\right]^n \left[\frac{sin\beta}{0.0896}\right]^m \qquad [8.4]$$

with n = 0.4 and m = 1.3 according to [MOO 89], where A_S is the contributive area upstream of each pixel S, and β is the slope in degrees.

SAGA function to calculate *LS* factor:
- Tools: *Hydrology > LS Factor, Field based...*

8.2.3. *Soil types and land cover factors*

8.2.3.1. *Soil erodibility: K factor*

Soil erodibility corresponds to susceptibility of a soil type to erosion. The soil loss potential is dependent on the substratum type, its cohesion, the size of the soil's particles and the pedogenesis.

Wischmeier *et al.* [WIS 71] propose a predictive methodology to estimate the K factor based on a nomograph[2] including the size of the particles, the organic matter content of the soil, the soil texture and permeability. This nomograph is derived from the following equation:

$$K = 2.1 \times M^{1.14} \times 10^{-6} \times (12 - MO) + 0.0325 \times (b - 2) + 0.025 \times (c - 3) \quad [8.5]$$

with:

– M is the sum of silt and fine sand in percent;

– MO is the organic matter in percent;

– b is soil texture;

– c is the permeability rate of the soil.

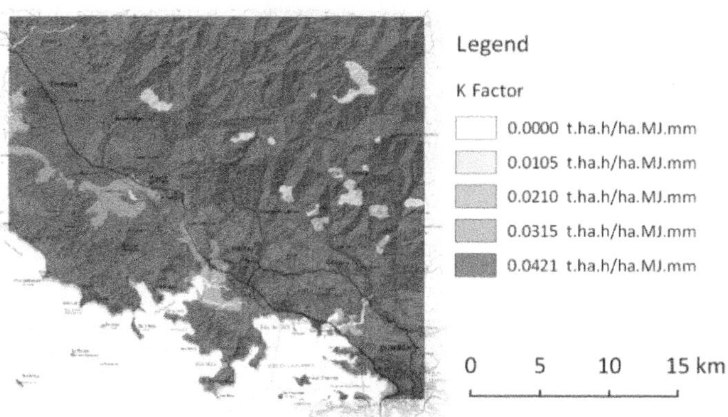

Figure 8.5. *K factor describing soil erodibility in t·ha·h·(ha·MJ·mm)$^{-1}$ over the region of Dumbéa — New Caledonia (166.3478°W, 22.0652°S). For a color version of the figure, see www.iste.co.uk/baghdadi/qgis4.zip*

2 A nomograph is a graphical tool consisting of a network of curves where each curve corresponds to a parameter. This makes it possible to find a value without calculating it but through a graphical estimate.

Without field experimentations to measure soil loss using rainfall simulations or through estimation using the method proposed by [WIS 78], empirical values are commonly used based on the table proposed by [STO 11]. These values are linked to soils types, from a soil map or a geological map, by performing attribute association of the K value according to the soil attribute (Figure 8.5).

QGIS function to create a new attribute based on an existing one:
- New attribute calculation: *Processing toolbox>QGIS Geoalgorithms>Vector table tools> Field calculator...*

8.2.3.2. *Vegetation cover: C factor*

C factor introduces the effect of fixation of soil particles by vegetation [WIS 78] as well as the effect of vegetation and crops cycles. C factor was first measured as the ratio of the soil loss of cropland under defined conditions to the corresponding soil loss on similar cropland under continuous fallow conditions [REN 97]. This factor varies over the interval [0; 1] (Figure 8.6).

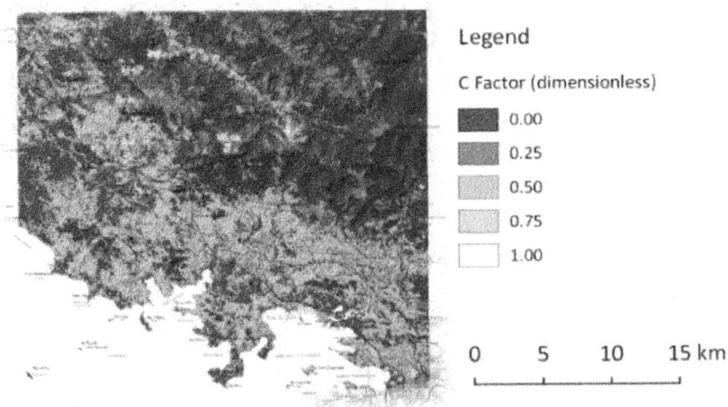

Figure 8.6. *C factor of the vegetation cover over the area of Dumbéa — New Caledonia (166.3478°W, 22.0652°S). For a color version of the figure, see www.iste.co.uk/baghdadi/qgis4.zip*

To take into account natural environments, the C factor is likened to the degree of soil cover by vegetation. The variation of the soil loss potential of a defined soil under different vegetation covers is then better adjusted (from bare soil to soil under dense vegetation) [REN 97].

Several methods exist to evaluate the C factor exploiting remote sensing imagery and calculation of a vegetation index (NDVI, vegetation fraction) or through the estimate of the mean vegetation cover according to a land cover map. This last method is presented here. It consists of associating an estimated value of vegetation cover for each class of land use.

QGIS function to create a new attribute based on an existing one:
- New attribute calculation: *Processing toolbox>QGIS Geoalgorithms>Vector table tools> Field calculator...*

8.2.3.3. *Soil conservation practices: P factor*

The P factor corresponds to conservation practices and represents the effect of specific cultivation practices whose aim is to reduce the quantities and the speed of runoff to limit the effects of erosion.

This factor is defined as the ratio of soil losses according to a given conservation practice to the soil losses of a crop arranged in rows aligned with the slope. It varies between [0; 1]. It is commonly defined from a correspondence table (Table 8.1) including the cultivation practices.

Conservation practices	P factor
Up and down slope	1.0
Cross slope	0.75
Contour farming	0.50
Strip cropping, cross slope	0.37
Strip cropping, contour	0.25

Table 8.1. *Example of correspondence table for the P factor [STO 11]*

P factor estimates rely on the knowledge level and the availability of sufficient databases concerning cultivation practices of the area of interest. Under natural conditions, the P factor is set to 1 [DUM 10a, PAY 12].

QGIS function to create a new attribute based on an existing one:
- New attribute calculation: *Processing toolbox>QGIS Geoalgorithms>Vector table tools> Field calculator...*

8.2.4. *Estimation of soil losses A*

The estimates of soil losses A are performed using equation [8.1] and the previously mentioned parameters. The A unit depends on the R and K factors' units, factors L, S, C and P being dimensionless.

Therefore the formula [8.3] determines the R factor in $\text{MJ}\cdot\text{mm}\cdot(\text{ha}\cdot\text{h}\cdot\text{year})^{-1}$. Depending on the data available or the desired evaluation, the K factor unit can vary. This factor is expressed as {([mass]·[area]·[period])/([area]·[energy]·[precipitations])}. It is important to ensure that:

– the period unit of the K factor conforms with the period unit of the R factor denominator;

– the energy unit of the K factor is the same as that of the R factor.

If this is not the case, the conversion of these two factors' units is necessary. In this case study, we use a K factor expressed in (t·ha·year)/(ha·MJ·mm). Thus, soil losses A is expressed in $\text{t}\cdot\text{ha}^{-1}\cdot\text{year}^{-1}$.

QGIS function to change R and K factors units:
- Raster calculator... : *Raster > Raster calculator...*

QGIS function to change the A factor units:
- Raster calculator... : *Raster > Raster calculator...*

8.2.5. *Limits of the method considered*

The use of the RUSLE equation to model erosion must take into account some limits inherent to the methodology and to each parameter:

– the choice of the RUSLE equation is often driven by the availability of the spatial parameters required over a selected region;

– erosion modeling using RUSLE provides soil losses estimates in t/ha/year. If erosion modeling relies on scientifically uncalibrated parameters or if the model results are not validated by comparison with field survey measurements, it is better to use the results as a relative or qualitative assessment of erosion risk (low to high erosion);

– access to certain modeling parameters, such as the P factor, may turn out to be difficult;

– the parameters used (soil map, topography, land use) are often available at different scales and precisions that could lead to inaccuracies related to the least precise spatial parameters;

– equation [8.1] considers each parameter with the same weight, whereas some parameters may have greater influence than others according to local conditions.

However, RUSLE modeling shows a good correlation between the location of erosion zones compared to field observations. The modeling allows also for comparison of different sites as well as temporal monitoring.

8.3. Implementation of the RUSLE model

This section presents a practical implementation of the RUSLE model on the Dumbéa area in New Caledonia (South-West Pacific). This area is characterized by a mountain chain reaching an altitude of 1,200 m and subject to heavy erosive processes due to runoff. The mountain range is bordered by a piedmont glacis followed by a coastal plain on the West Great Lagoon of the Grande Terre of New Caledonia.

8.3.1. *Software and data*

8.3.1.1. *Software*

Raster processing and vector analysis performed in this implementation are made with the OSGeo4W software suite that includes QGIS[3] version 2.18 and SAGA[4] GIS version 2.3.1.1. (long-term release) [LAC 17].

The OSGeo4W setup program can be downloaded at the following link: https://www.qgis.org/fr/site/forusers/download.html.

Figure 8.7. *QGIS v2.18 and SAGA GIS v2.3.1.1. software*

Data analysis can be performed using other similar software such as ArcGIS, Idrisi, GRASS or ERDAS Imagine.

3 QGIS is user-friendly GIS software distributed under GNU public licence. It's an official project of the Open Source Geospatial foundation (OSGeo) [QGI 17].

4 SAGA is a open source GIS software SIG distributed under GPL licence dedicated to geoscientific analyses [CON 15].

8.3.1.2. *Data input*

The K factor vector file is provided for this implementation (facteur_k_dumbea.shp). This file comes from the soil map of New Caledonia [LAT 78] where k values are estimated according to the k table from [DUM 10a]. The reference coordinate system is RGNC91-93, Lambert NC projection.

Climate data are an extraction of the global precipitation model provided by *World Climate Rainfall Data* [HIJ 05].

Other datasets used in this implementation can be downloaded from the New Caledonia geographical web server: http://georep.nc/. Data are distributed under a *creative commons* license (the use conditions for each dataset are described in their metadata):

– DEM of New Caledonia at 10 m resolution;

– New Caledonia land cover database, realized in 2008. This classification, adapted to the context of New Caledonia, consists of 19 classes grouped into four categories. This classification results from a semiautomatic procedure of satellite image analysis. Its validity is therefore not absolute. The classification was validated with a kappa coefficient of 75.5% [DTS 08].

8.3.1.2.1. Downloading World Climate Rainfall data

The data can be downloaded from the official WorldClim Website. The file to download is the precipitation (mm) file at 30-s resolution.

File name: wc2.0_30s_prec.zip.

Link: http://worldclim.org/version2.

File size: 978 Mb.

8.3.1.2.2. Downloading DEM

The DEM can be downloaded from the geographic web server of New Caledonia. The file is the terrestrial DEM at 10m resolution in TIF format within the reference coordinate system RGNC91-93.

File name: mnt10NC_RGNC9193_LAMBERTNC.zip.

Link: http://www.geoportal.gouv.nc/geoportal/catalog/search/resource/details.page?uuid=%7B210E58EC-B775-43EC-AEB9-55738689244B%7D.

File size: 108 Mb.

8.3.1.2.3. Downloading Land cover

The land cover classification is the result of a semiautomatic processing procedure using SPOT satellite imagery for the period 2006–2008. Nineteen classes are grouped into four categories: water and mangroves, urban area, vegetation on volcanic and sedimentary rocks, vegetation on ultrabasic rocks [DTS 08]. The land cover database is delivered in shapefile format within the reference coordinate system RGNC91-93.

File name: occupation_du_sol_2008_SPOT5_approche_objet_shapefile.zip.

Link: http://www.geoportal.gouv.nc/geoportal/catalog/search/resource/details.page?uuid=%7B44E58684-49BA-4F60-AFB7-BB24464FF4B4%7D.

File size: 71.6 Mb.

8.3.2. *Step 1: R factor calculation*

The compressed WorldClim file contains 12 GeoTIF files named wc2.0_30s_prec_m with m a number between 1 and 12. These files correspond to mean monthly precipitation .

These files cover the entire world and the reference coordinate system is WGS84. The projection of these files, being different from the other dataset, needs to be reprojected.

8.3.2.1. *Calculation of the average annual rainfall P*

The average annual rainfall corresponds to the amount of precipitation in mm received on average in each mesh of the model. In this implementation, the average annual rainfall is estimated by summing the mean monthly precipitation according to the formula:

$$P = \sum_{i=1}^{12} P_i \qquad [8.6]$$

with P_i the mean precipitation of the month i.

<table>
<tr><th>Approach</th><th>Practical work in QGIS</th></tr>
<tr><td>1. Opening of the 12 mean monthly precipitation files</td><td>Using QGIS:
• Open the raster files :
o wc2.0_30s_prec_01.tif
o wc2.0_30s_prec_02.tif
o ...
o wc2.0_30s_prec_12.tif</td></tr>
<tr><td>2. File verification</td><td>Using QGIS:
• Visually verify the display of the data (overlay, rendering)
• If necessary, check the metadata of each layer by selecting the layer, right-clicking and selecting properties and then the tab</td></tr>
<tr><td>3. Average annual rainfall calculation</td><td>In the menu bar:
• Click on: Raster > Raster Calculator...
In the Raster Calculator:
• Enter the following expression:
"wc2.0_30s_prec_01@1"+"wc2.0_30s_prec_02@1"+
"wc2.0_30s_prec_03@1"+"wc2.0_30s_prec_04@1"+
"wc2.0_30s_prec_05@1"+"wc2.0_30s_prec_06@1"+
"wc2.0_30s_prec_07@1"+"wc2.0_30s_prec_08@1"+
"wc2.0_30s_prec_09@1"+"wc2.0_30s_prec_10@1"+
"wc2.0_30s_prec_11@1"+"wc2.0_30s_prec_12@1"
• Save the file as Worldclim_p_annuelles_4326.tif
The result is a raster file with pixel values ranging from 0 to 11,256.
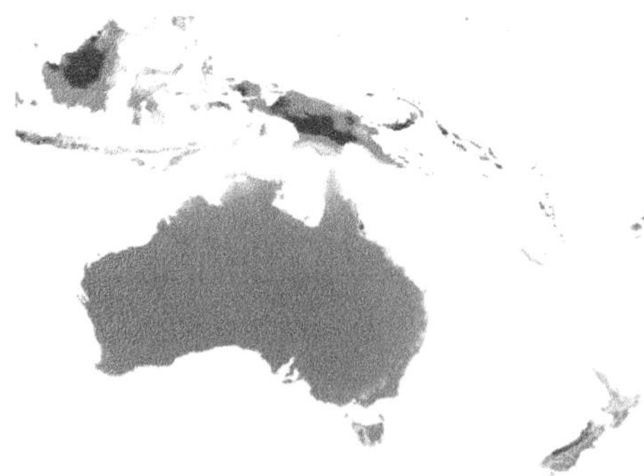
For a color version of the figure above, see www.iste.co.uk/baghdadi/qgis4.zip
The calculation may take a few minutes.</td></tr>
</table>

8.3.2.2. *Average annual rainfall reprojection and clipping*

The average annual rainfall is calculated over the entire world in the reference coordinate system WGS84. The file size may limit calculation performance. And the reference coordinate system is different from the other datasets used. Indeed, New Caledonia datasets are in the RGNC91-93 reference coordinate system (EPSG:3163).

Therefore, average annual rainfall should be reprojected and clipped on the area covered by the other datasets. The geographic intersection of all the datasets corresponds to the shapefile layer "facteur_k_dumbea.shp" that covers the most restricted area.

Approach	Practical work in QGIS

1. File verification	Using **QGIS**:
	• Open the previously created file:
	  o `Worldclim_p_annuelles_4326.tif`
	• Open the shapefile:
	  o `facteur_k_dumbea.shp`
	• Make sur that the QGIS on-the-fly reprojection function is enable to view layers' overlay.

2. Creation of a clipping mask	In the **menu bar**: • Click on: **Vector>geoprocessing tools> Convex Hull(s)...** • Choose `facteur_k_dumbea.shp` in input • Leave the Field option as default [Not set] • Choose the method "Create a single minimum convex hull" • For "Convex Hull", indicate the output name masque.shp • Click OK

<table>
<tr>
<td></td>
<td>
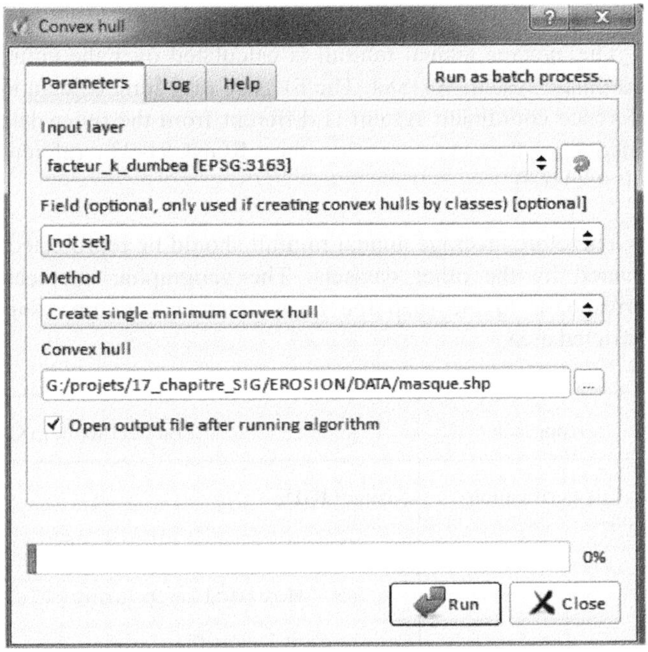

- Check the created file
</td>
</tr>
<tr>
<td>3. Clipping of the GeoTIF file</td>
<td>
In the menu bar:

- Click on: **Raster >Extraction>Clipper…**
- Choose the file `Worldclim_p_annuelles_4326.tif` as input
- Enter the output file `Worldclim_p_annuelles_4326_clip.tif`
- In the section **Clipping mode**, choose **Mask layer** and set the layer `masque.shp` created previously
- Choose the option **Crop the extent of the target dataset to the extent of the cutline**
- And select the option **Keep resolution of input raster**
</td>
</tr>
</table>

<table>
<tr>
<td></td>
<td>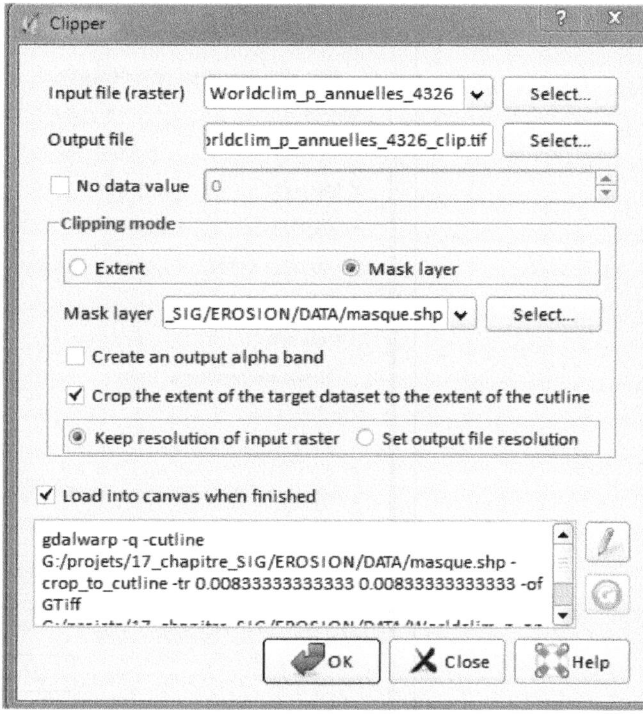

<ul><li>Check the created file</li></ul></td>
</tr>
<tr>
<td>4. Reprojection of the TIF file</td>
<td>In the menu bar:
<ul><li>Click on: Raster >Projections>Warp (Reproject)...</li></ul>
In the Warp window:
<ul>
<li>Select the file <code>Worldclim_p_annuelles_4326_clip.tif</code> as input;</li>
<li>Enter the output file name <code>Worldclim_p_annuelles_3163_clip.tif</code></li>
<li>Check the source SRS - EPSG:4326</li>
<li>Select Target SCR and choose the EPSG:3163</li>
<li>Check that the resampling method is set to "Near"</li>
</ul></td>
</tr>
</table>

	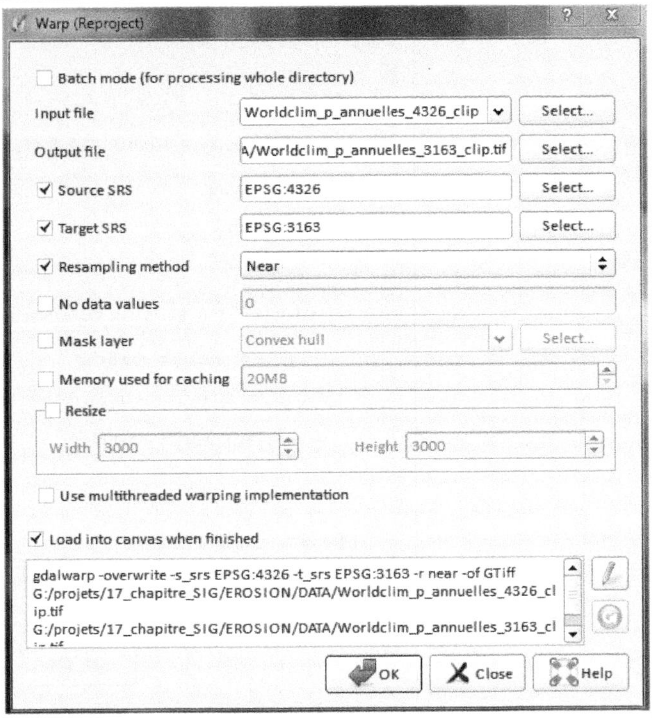 • Click OK • Check the created file.

8.3.2.3. *Rainfall aggressiveness R factor calculation*

The rainfall aggressiveness R factor is calculated according to the formula [8.3], i.e. from the average annual rainfall that is reprojected and clipped over the area of interest.

Approach	Practical work in QGIS
1. File verification	In **QGIS**: • Open the previously created file: o `Worldclim_p_annuelles_3163_clip.tif`
2. *R* factor calculation	In the **menu bar**: • Click on: **Raster > Raster Calculator…**

<table>
<tr><td></td><td>In the Raster Calculator:
• Enter the following expression:

```
1.73 *
"Worldclim_p_annuelles_3163_clip@1"*0.5
```

• Save the file as <code>facteur_r_3163.tif</code>
The result is a raster file with values ranging between 953 and 2,383.
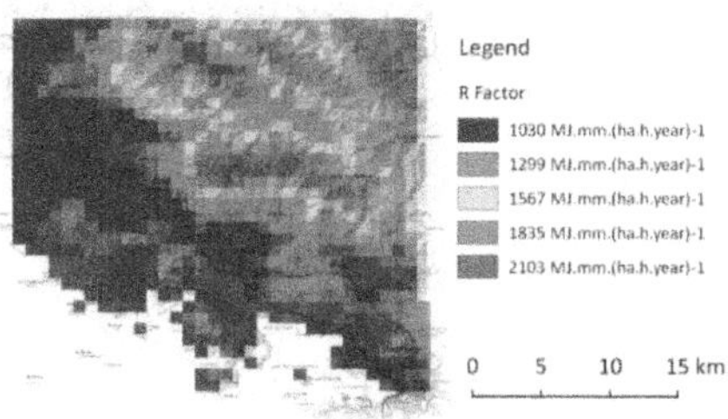

For a color version of the figure above, see www.iste.co.uk/ baghdadi/qgis4.zip</td></tr>
</table>

8.3.3. *Step 2: LS factor calculation*

LS factor calculation is performed using SAGA GIS software. In order to optimize the processing of the data, the DEM is clipped in advance using QGIS.

8.3.3.1. *DEM clipping over the area of interest*

The DEM is clipped using the same procedure as that explained in section 8.3.2.2. However the DEM is not reprojected because it is already in the RGNC91-93 reference coordinate system.

Approach	Practical work in QGIS
1. File verification	In **QGIS**: • Open the file: o `mnt10_rgnclambert.tif` • Open the shapefile: o `masque.shp` • Make sur that the QGIS on-the-fly reprojection function is enabled to view layers' overlay.

2. Clipping of the GeoTIF file F	In the **menu bar**: • Click on: **Raster >Extraction>Clipper…** Choose the `file mnt10_rgnclambert.tif` as input. • Enter the output file name `mnt10_rgnclambert_clip.tif` • In the section **Clipping mode**, choose **Mask layer** and set the layer masque.shp created previously • Choose the option **Crop the extent of the target dataset to the extent of the cutline** • And select the option **Keep resolution of input raster** • Check the created file

8.3.3.2. *SAGA GIS interface*

To calculate the LS factor, we need to start SAGA GIS from the computer's start menu.

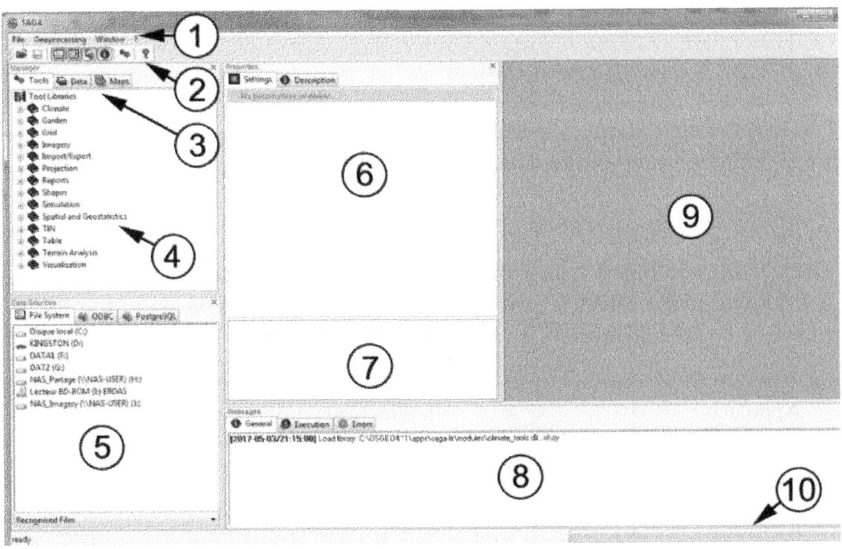

Figure 8.8. *SAGA GIS Interface*

The SAGA interface (Figure 8.8) is made up of:

1) the menu bar;

2) an icons bar to manage the interface (from left to right: open a file; save a project; hide the management window; hide the properties window; hide the data source window; hide the message console; find and launch an algorithm, help);

3) selection tab of the management window: tools, data and map;

4) management window: this window gives access to the management functionalities of each section of the management window (toolbox, opened data management, map management);

5) data source management window: three types of access are proposed using three different tabs: by browsing the file tree, using an ODBC connection to databases, or by connecting a PostgresSQL database;

6) the properties window: this makes it possible to manage tools, data and map properties. Depending on whether a tool or a data/map is in use, different tabs are available;

7) algorithm management window: this makes it possible to tune algorithms' parameters, restore the default settings, launch the process and save a specific setting;

8) the messages console: this displays messages and application logs under the General, Execution and Errors tabs;

9) map window, to visualize maps;

10) progress bar.

Like GRASS, SAGA groups the data according to their boundaries and resolution. In this way, the data imported into the software will appear in the data tab of the management window as:

Data
 └ Grids
  └ Pixel size; number of pixels x,y; coordinates SW of the Grid
  (ex: 10; 3000x 3000y; 420286.925218x 219397.144291y)

If two rasters of different sizes are imported, they will appear under two different grids. Furthermore, the processing between two rasters in SAGA requires that they belong to the same Grid and therefore have the same resolution and the same extend.

8.3.3.3. *Import DEM into SAGA*

Importing of the previously clipped DEM to the area of interest is performed using the quick data access procedure of the data source management window.

<table>
<tr><th>Approach</th><th>Practical work in QGIS</th></tr>
<tr><td>1. Quick data import</td><td>

In **SAGA**:

- In the data access window, choose the DEM clipped on the area of interest:
    - `mnt10_rgnclambert_clip.tif`

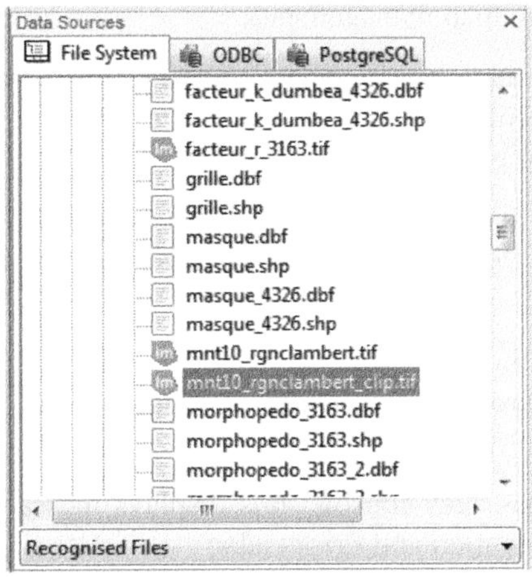

- Double-click the file
- Verify the successful import of the file on the data tab of the management window

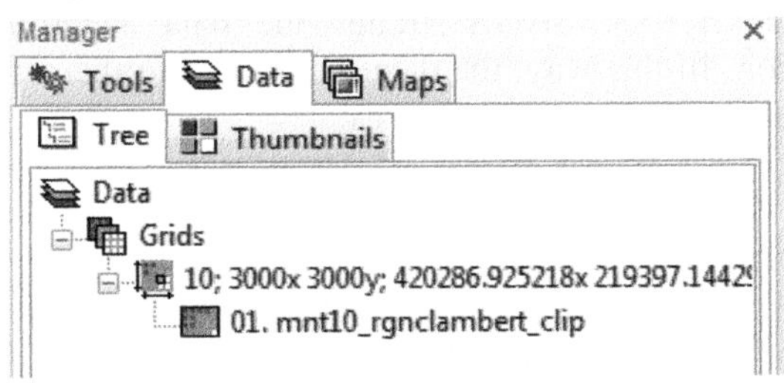

- To view the data, double-click the data in the **data tab** of the **management window**. The selected data is displayed in the **map window**. Its metadata, import parameters, legend and usage history are displayed in the **properties window**.

</td></tr>
</table>

	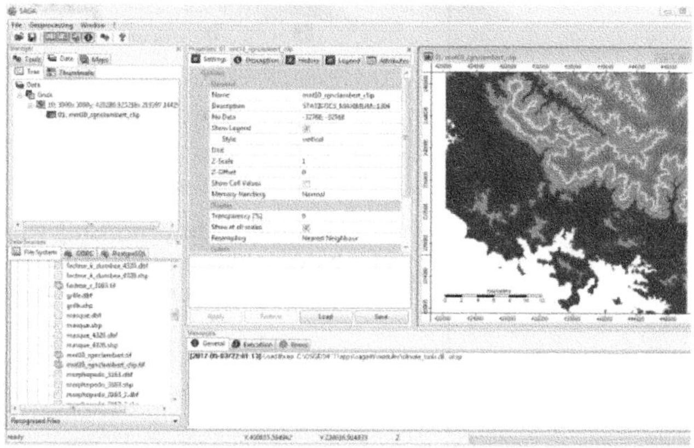 • Check the resampling method used to import the DEM in the **history tab**. By default, the resampling setting is set to B-Spline Interpolation.

SAGA is a software dedicated to GIS processing in geosciences. It therefore propose a wide range of parameters for each operation in order to control the integrity of the processed data. The B-Spline interpolation can introduce unwanted deformations in elevation values of the DEM. So we will prefer to have the choice of the interpolation used when we import DEM in SAGA.

Approach	Practical work in QGIS
1. Import with specific settings	In **SAGA**: • Close the previously imported DEM by selecting Grids on the data tab, right-click and choose close. The software asks if you want to delete the data. Choose yes. • In the management window, in the tool tabs, find the functionality **Import/Export > GDAL/OGR > Import Raster** • By selecting Import Raster, the settings window of the properties window indicates the properties of the import function • In the **File** section, choose the file `mnt10_rgnclambert_clip.tif` • In the **Resampling** section, select the option **Bilinear Interpolation**

	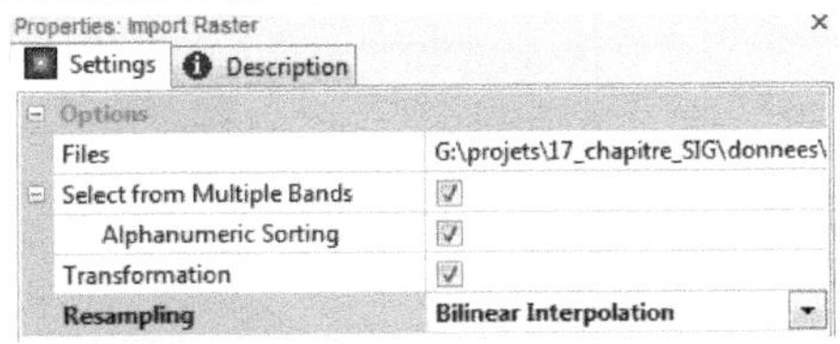 • Click **Apply** in the **processing window** (the settings changes in bold in the settings tabs of the properties window disappear) and then click **Execute** 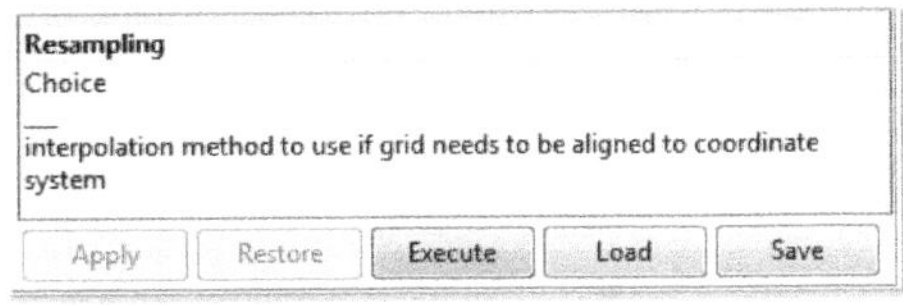 • **The messages console** indicates your actions' log.  • Import is successful if the log indicates *Tool execution succeeded*

8.3.3.4. *LS factor calculation*

The LS factor is a parameter derived from DEM. It is exploited in hydrological and topographical analyses. In SAGA, algorithms are grouped thematically in the Tools tab of the management window.

Approach	Practical work in QGIS
1. LS factor algorithm settings	In **SAGA**: • Go to the Tools tab of the management window • Go to **Terrain Analysis>Hydrology** and select **LS-Factor, Field based** algorithm • In the properties window, Description tab, a complete description of the algorithm is provided including the author, a clear description, bibliography and a short explanation of all the settings • In the properties window, Settings tab, choose the Grid system corresponding to the previously imported DEM • Choose the layer `mnt10_rgnclambert_clip` in input of >>Elevation

<table>
<tr>
<td></td>
<td>

- Leave the settings Upslope Length Factor, Effective Flow Length and UpslopeSlope by default to <not set>
- Check that the output << LS Factor is set to <create>
- Leave the settings Fields and Sediment balance by default to <not set>
- Check that the LS Calculation method is Moore & Nieber 1989 which corresponds to equation [8.4].
- Leave the options Type of Slope, Specific Catchment Area and Stop at Edge as default

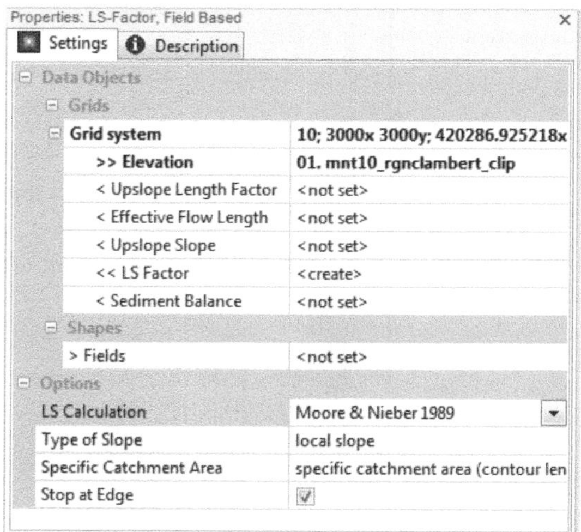

</td>
</tr>
<tr>
<td>2. LS processing and results verification</td>
<td>

- Click **Apply** in the **process management window** (changes in bold in the Settings tab of the properties window disappear) and then click **Execute**

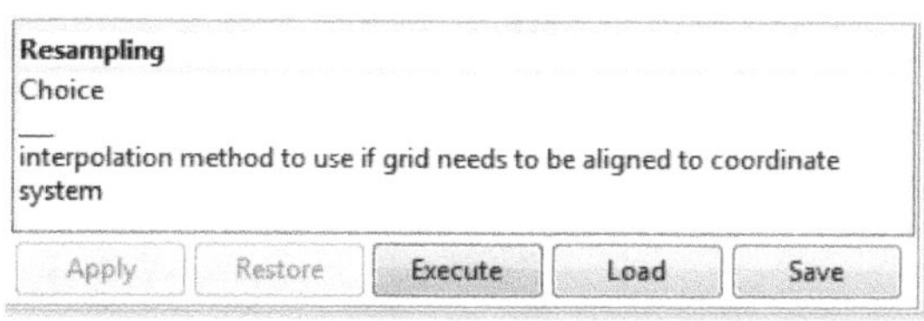

- **The message console** shows the log of your actions

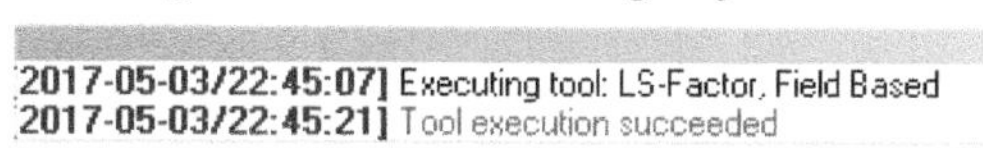

- LS calculation has succeeded if the log indicates *Tool execution succeeded*
- To view the result, double-click on the layer created in the **data tab** of the **management window**. Note in the data tab

</td>
</tr>
</table>

that the computed LS factor is added as a new layer in the DEM Grids

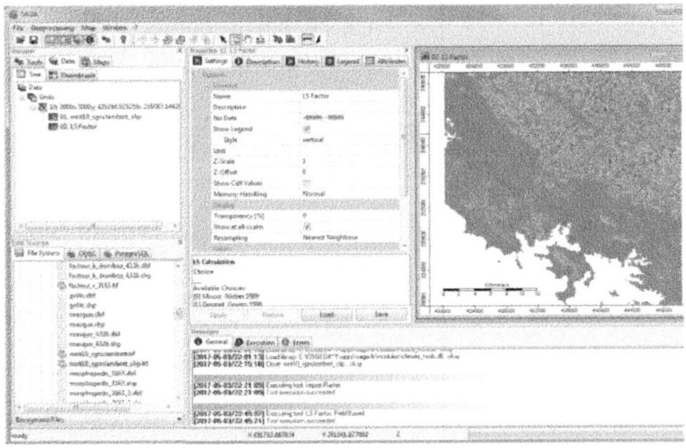

- Verify the **resampling** in the **history tab**. It is set by default as B-Spline Interpolation
- The result is a raster with values ranging from 3.2×10^{-7} to 355.96

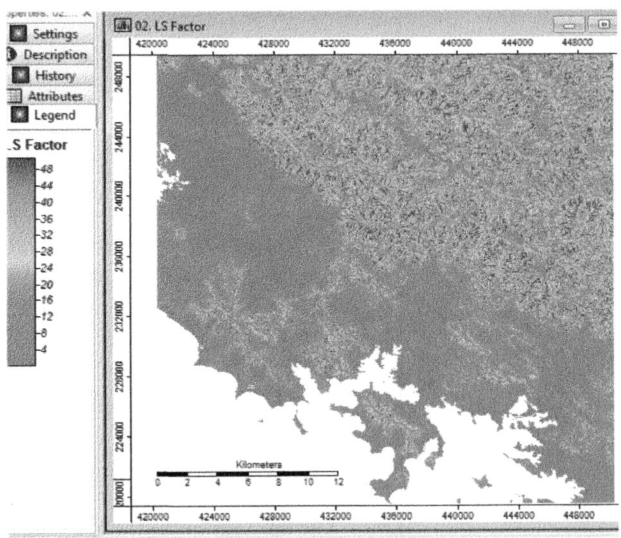

For a color version of the figure above, see www.iste.co.uk/baghdadi/qgis4.zip

- When viewing the data, color stretching is automatically adjusted according to the raster values' histogram. In this example, the majority of the LS values range between 0 and 50

8.3.3.5. *Export data from SAGA*

SAGA creates temporary layers when executing algorithms. Layers are saved when you save the project or when you want to export the data to other software. The subsequent calculation of soil losses using the RUSLE model is done in QGIS, so the LS factor is exported in a QGIS compliant format.

Approach	Practical work in QGIS
1. Settings of data export	In **SAGA**: • In the **management window, Tools tab**, go to **Import/Export> GDAL/OGR** and select Export GeoTIFF • In the properties window, Settings tab o Select the Grid system corresponding to the DEM o Specify outputs as **>>Grid(s)** by clicking on the **...** button on the right 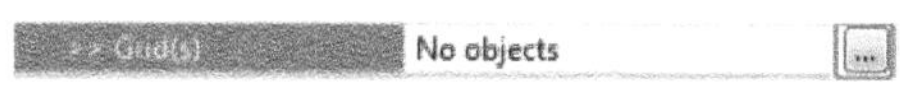 o In the selection window that opens, select 02.LS Factor in the left side and pass it in the right side with the > button. o Click OK 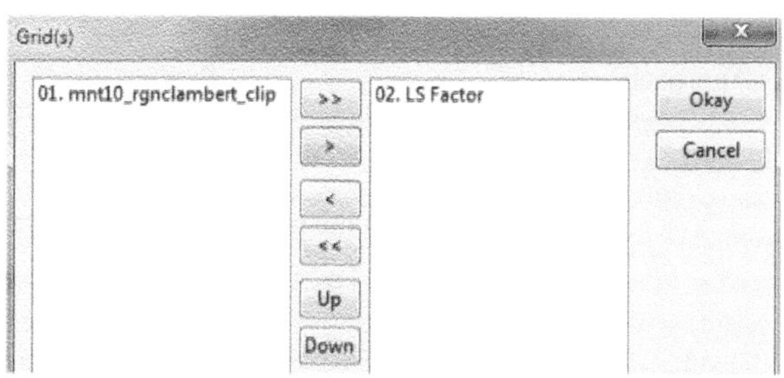 • Specify the export file name in File: facteur_LS_3163.tif 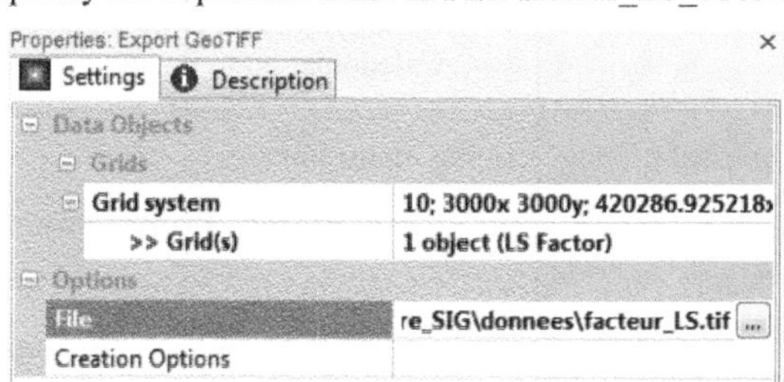

2. Export and result check	• Click **Apply** on the **process management window** (changes in bold in the Settings tab of the properties window disappear) and then click **Execute** 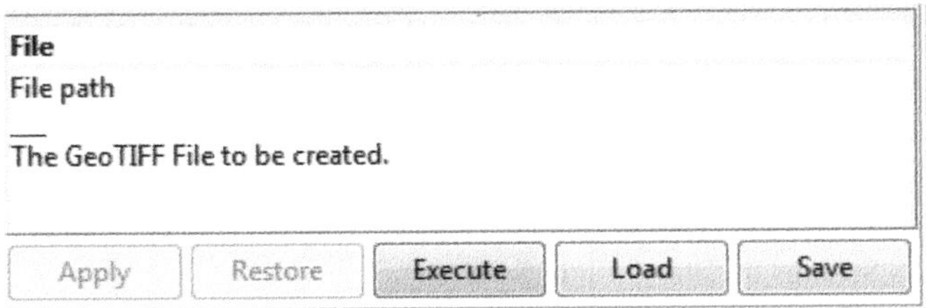 • **The messages console** indicates your actions' log.  • Export is successful if the log indicates *Tool execution succeeded*

The SAGA software does not include all the QGIS projection libraries. Also, the projection of the exported file may not be well recognized by QGIS, which is the case for CRS RGNC91-93. The projection must therefore be redefined in QGIS for the LS file exported from SAGA.

Approach	Practical work in QGIS
1. File verification	In **QGIS**: • Open the previously created file: o `facteur_LS_3163.tif` • In the file metadata (properties, tab), check that the CRS is (USER:10000...), then go to 2. • if the projection is EPSG:3163 - RGNC91-93 / Lambert New Caledonia, go to the next step (8.3.4.).
2. Assignment of RGNC91-93/ Lambert New Caledonia projection	In the **menu bar**: • Click: **Raster >Projections > Assign projection…** • Choose the file `facteur_LS_3163.tif` • Choose the following CRS EPSG:3163

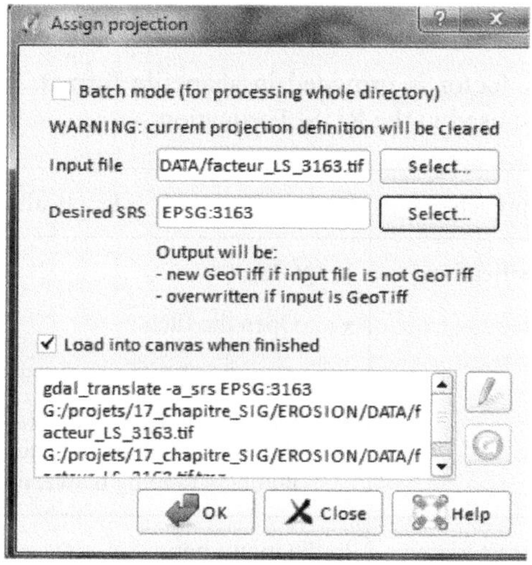

- Click OK
- When the process is complete, a second `facteur_LS_3163.tif` file appears in the layer list. Close both files
- QGIS had created a `facteur_LS_3163.tmp` file in the same folder as the source file. Delete the `facteur_LS_3163.tif` and replace the tmp extension of the `facteur_LS_3163.tmp` file with tif
- Open this new file:
  - o `facteur_LS_3163.tif`
- Check the projection in the data properties window

The result is a raster with values ranging from 0 to 356.

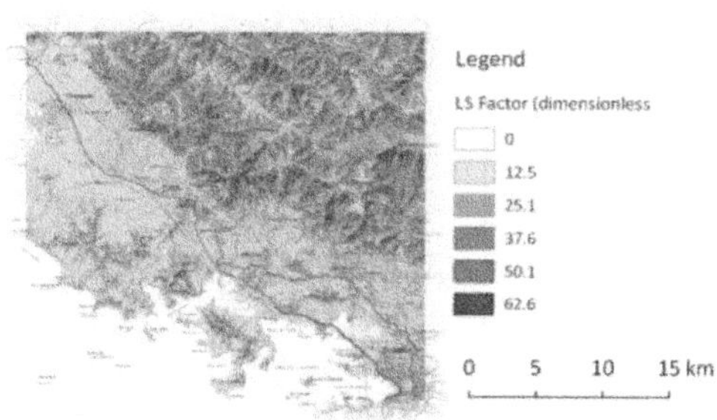

For a color version of the figure above, see www.iste.co.uk/baghdadi/qgis4.zip

8.3.4. *Step 3: preparation of the K factor*

The K factor is provided in shapefile format. The vector must be rasterized in order to be used in the RUSLE equation.

Approach	Practical work in QGIS
1. File verification	In **QGIS**: • Open the file: o `facteur_k_dumbea.shp` • Open the attribute table (select the layer, right-click > **Open Attribute Table**) and check that the k column contains real numbers ranging between 0 and 0.0421
2. Vector rasterizing	In the **menu bar**: • Click on: **Raster >Conversion > Rasterize (vector to raster)...** • Select the input file `facteur_k_dumbea.shp` • Select the k column as Attribute field • Specify the output file `facteur_k_3163.tif`, and click OK • A pop-up window opens indicating that the output file does not exist. You must set the file size or the resolution to be able to create it. Click OK • Select the Resolution radio button expressed in map units per pixel and specify 10.000 in horizontal and vertical 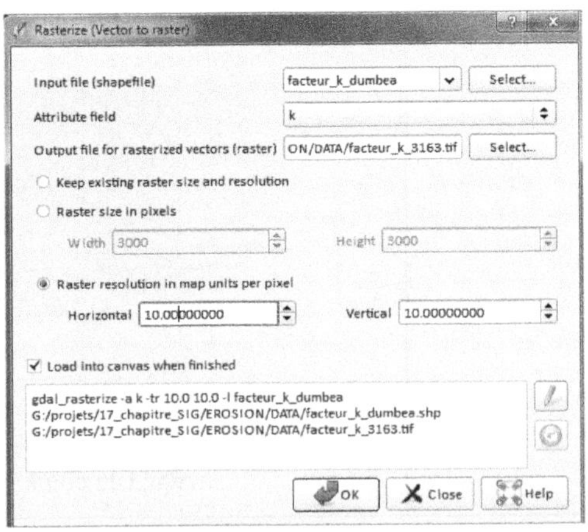

- Click OK
- Check the minimum and maximum values of the created raster which must be 0 and 0.0421, respectively, in the layer properties window, metadata tab

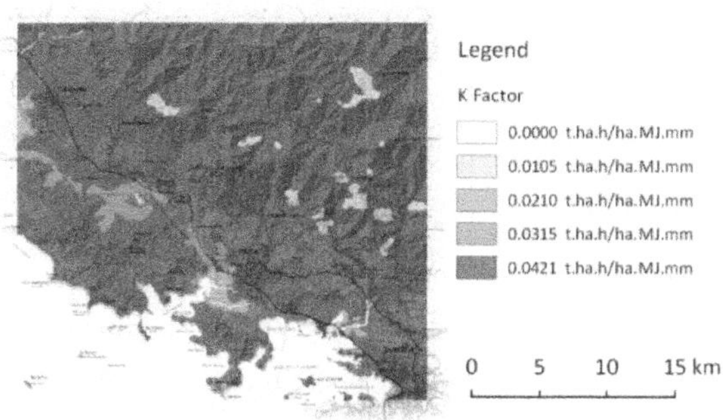

For a color version of the figure above, see www.iste.co.uk/baghdadi/qgis4.zip

8.3.5. *Step 4: C factor creation*

C factor is estimated here by the association of a C value to each class of land use. The chosen values are those proposed by [DUM 10b] and presented in Table 8.2.

Land cover type	C factor
Forest and near forest vegetation, dense mangroves	0.001
Savana	0.04
Shrublands on ultrabasic soils	0.25
Swamp	0.28
Bush and open vegetation	0.72
Bare soils	1
Water	1

Table 8.2. *Land cover types and associated C factor values [DUM 10b]*

8.3.5.1. *Vector file clipping*

The land cover database concerns the entire territory of New Caledonia. The first step consists of clipping the file over the same area as the *K* factor.

Approach	Practical work in QGIS
1. File verification	In **QGIS**: • Open the files: o `occupation_sol_2008_spot5_objet.shp` o `masque.shp` created in section 8.3.2.2.
2. Land cover database clipping	In the **menu bar**: • Go to: **Vector>Geoprocessing tools> Clip...** • Choose the input vector file `occupation_sol_2008_spot5_objet.shp` • Choose the clip layer `masque.shp` • Make sure that the two options "Use only selected features" are not selected • Specify the output shapefile name: o `occupation_sol_2008_spot5_objet_clip.shp` • Click OK • Check the file created

8.3.5.2. *Updating the attribute table*

This update consists of creating a new field corresponding to the C factor values. This field is populated with C values that correspond to each class of Table 8.2.

Approach	Practical work in QGIS
1. Correct character problems	Characters such as ' cause problems when applying conditions to create the new attribute. They must be corrected. • Open the attribute table of the vector layer `occupation_sol_2008_spot5_objet_clip.shp` In the **Attribute table**, • Select all the polygons whose field CLASSE = *Zones d'habitation* by clicking on 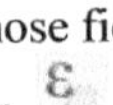 o In the field **Expression**, enter the following condition: "GRIDCODE" = 24 o Click **Select**

	• Open the Field Calculator
	  o Select **Update existing field**
	  o Select the field **CLASSE**
	  o Verify that the option **Only update XXX selected features** is checked
	  o In the field **Expression**, add the new attribute as follow:
	'Zone d habitation'
	  o Click OK
	  o Deselect the objects by clicking

2. Updating the attribute table	• Open the Field Calculator in clicking
	In the **Field calculator**:
	• Select **Create a new field**
	• Enter the **field name**: `C`
	• Define the field **Type**: `decimal number (real)`
	• Set the **Output field lenght** = `12` (total lenght of the number)
	• Set the **Precision** = `6` (number of digits after the decimal point)
	• In the field **Expression**, enter the following condition:
	`CASE WHEN "CLASSE" ='Eau douce' THEN 1`
	`WHEN "CLASSE" ='Eau marine' THEN 1`
	`WHEN "CLASSE" ='Forêt sur substratultramafique' THEN 0.001`
	`WHEN "CLASSE" ='Forêt sur substrat volcano-sédimentaire' THEN 0.001`
	`WHEN "CLASSE" ='Mangrove clairsemée' THEN 0.28`
	`WHEN "CLASSE" ='Mangrove dense' THEN 0.001`
	`WHEN "CLASSE" ='Maquis dense paraforestier' THEN 0.001`
	`WHEN "CLASSE" ='Maquisligno-herbacé' THEN 0.25`
	`WHEN "CLASSE" ='Nuages' THEN 0`
	`WHEN "CLASSE" ='Savane' THEN 0.04`
	`WHEN "CLASSE" ='Sol nu sur substrat ultramafique' THEN 1`
	`WHEN "CLASSE" ='Sol nu sur substrat volcano-sédimentaire' THEN 1`
	`WHEN "CLASSE" ='Tanne' THEN 1`
	`WHEN "CLASSE" ='Végétation arbustive sur`

	substrat volcano-sédimentaire' THEN 0.72 WHEN "CLASSE" ='Végétation éparse sur substrat ultramafique' THEN 0.72 WHEN "CLASSE" ='Végétation éparse sur substrat volcano-sédimentaire' THEN 0.72 WHEN "CLASSE" ='Zones cultivées, labours' THEN 0.04 WHEN "CLASSE" ='Zoned habitation' THEN 1 WHEN "CLASSE" ='Zones sombres (non interprétables)' THEN 0 Else 0 End • Click OK

8.3.5.3. *Rasterizing*

In order to be used in the RUSLE equation, the C factor must be rasterized.

Approach	Practical work in QGIS
1. File verification	In **QGIS**: • Open the file: o occupation_sol_2008_spot5_objet_clip.shp • Open the attribute table (Select the layer, right-click > **Open attribute table**) and verify that the C column values range between 0 and 1
2. Vector file rasterizing	In the **menu bar**: • Go to: **Raster >Conversion > Rasterize (vector to raster)…** • Select the input file occupation_sol_2008_spot5_objet_clip.shp • Select the c column as Attribute field • Specify the output file facteur_c_3163.tif, and click OK • A pop-up window opens indicating that the output file does not exist. You must set the file size or the resolution to be able to create it. • Click OK • Select the Resolution radio button expressed in map units per

pixel and specify 10.000 in horizontal and vertical

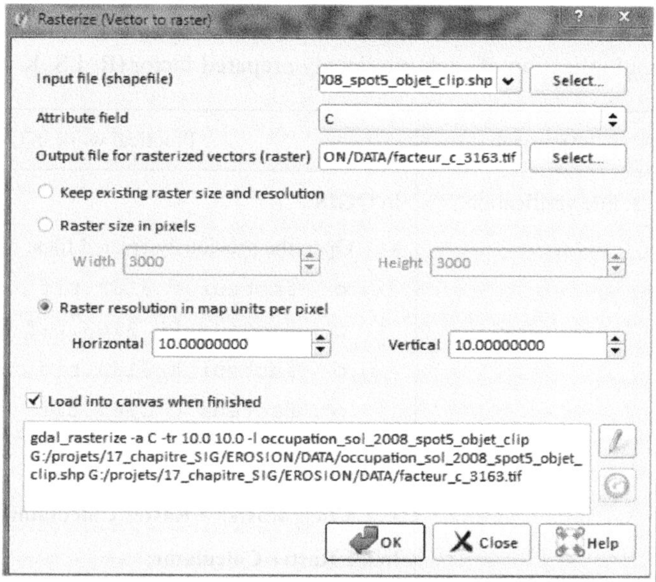

- Click OK
- Verify that the raster file minimum and maximum values are, respectively, 0 and 1 in the layer properties window, metadata tab

The result is a raster file with values ranging from 0 to 1.

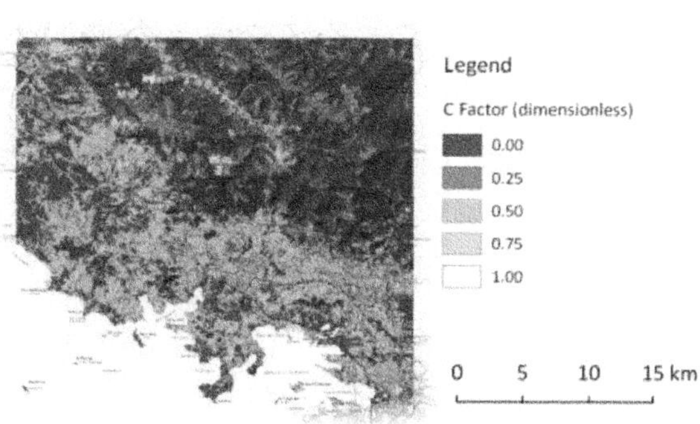

For a color version of the figure above, see www.iste.co.uk/baghdadi/qgis4.zip

8.3.6. *Step 5: soil loss A calculation from the RUSLE equation*

The calculation of soil losses A according to the formula [8.1] corresponds to the multiplication of each previously prepared factor (R, LS, K and C).

Approach	Practical work in QGIS
1. File verification	In **QGIS**: • Open the previously created files: o `facteur_r_3163.tif` o `facteur_LS_3163.tif` o `facteur_k_3163.tif` o `facteur_c_3163.tif`
2. A factor calculation	In the **menu bar**: • Click: **Raster > Raster Calculator...** In the **Raster Calculator**: • Enter the following expression: `"facteur_LS_3163.tif@1"*"facteur_c_3163@1"*" facteur_k_3163@1"*"facteur_r_3163@1"` • Save the file as `pertes_en_sol_A.tif` The result is a raster file with values range between 0 and 19,321 t/ha/year.  For a color version of the figure above, see www.iste.co.uk/baghdadi/qgis4.zip

8.4. Bibliography

[CON 15] CONRAD O., BECHTEL B., BOCK M., *et al.*, "System for Automated Geoscientific Analyses (SAGA) v. 2.1.4", *Geoscientific Model Development*, vol. 8, no. 7, pp. 1991–2007, 2015.

[DES 96] DESMET P.J.J., GOVERS G., "A GIS Procedure for Automatically Calculating the USLE LS Factor on Topographically Complex Landscape Units", *Journal of Soil and Water Conservation*, vol. 51, no. 5, pp. 427–433, 1996.

[DTS 08] DTSI, "Notice de La Classification de L'occupation Du Sol de La Nouvelle-Calédonie Par Approche Objet V1.0 - 2008", available at: http://sig-public.gouv.nc/Notice-Occupationdusol2008-SPOT5-approcheobjet.pdf, 2008.

[DUM 10a] DUMAS P., PRINTEMPS J., MANGEAS M. *et al.*, "Developing Erosion Models for Integrated Coastal Zone Management: A Case Study of The New Caledonia West Coast", *Marine Pollution Bulletin*, vol. 61, nos 7–12, pp. 519–529, 2010.

[DUM 10b] DUMAS P., PRINTEMPS J., "Assessment of Soil Erosion Using USLE Model and GIS for Integrated Watershed and Coastal Zone Management in the South Pacific Islands", *Proceedings Interpraevent, International Symposium in Pacific Rim, Taipei, Taiwan*, pp. 856–866, 2010.

[HIJ 05] HIJMANS R.J., CAMERON S.E., PARRA J.L. *et al.*, "Very High Resolution Interpolated Climate Surfaces for Global Land Areas", *International Journal of Climatology*, vol. 25, no. 15, pp. 1965–1978, 2005.

[LAC 17] LACAZE B., DUDEK J., PICARD J., "GRASS GIS Software with QGIS", in BAGHDADI N., MALLET C., ZRIBI M. (eds), *QGIS and Generic Tools*, ISTE Ltd, London and John Wiley & Sons, New York, 2017.

[LAT 78] LATHAM M., QUANTIN P., AUBERT G., Etude des sols de la Nouvelle-Calédonie: nouvel essai sur la classification, la caractérisation, la pédogenèse et les aptitudes des sols de Nouvelle-Calédonie : carte pédologique de la Nouvelle-Calédonie à 1/1 000 000: carte d'aptitudes culturale et forestière des sols de Nouvelle-Calédonie à 1/1 000 000, Notice Explicative 78, ORSTOM, Paris, 1978.

[MOO 89] MOORE I.D., NIEBER J.L., "Landscape Assessment of Soil Erosion and Nonpoint Source Pollution", *Journal of the Minnesota Academy of Science (USA)*, vol. 55, pp. 18–25, 1989.

[MOO 91] MOORE I.D., GRAYSON R.B., LADSON A.R., "Digital Terrain Modelling: A Review of Hydrological, Geomorphological, and Biological Applications", *Hydrological Processes*, vol. 5, no. 1, pp. 3–30, 1991.

[PAN 91] PANUSKA J.C., MOORE I.D., Water Quality Modeling: Terrain Analysis and the Agricultural Non-Point Source Pollution (AGNPS) Model, Water Resources Research Center, University of Minnesota, 1991.

[PAY 12] PAYET E., DUMAS P., PENNOBER G., "Modélisation de l'érosion hydrique des sols sur un bassin versant du sud-ouest de Madagascar, le Fiherenana", *VertigO*, vol. 11, no. 3, 2012.

[QGI 17] QGIS, "Découvrez QGIS", available at: https://www.qgis.org/fr/site/about/index.html, 2017.

[REN 94] RENARD K.G., FREIMUND J.R., "Using Monthly Precipitation Data to Estimate the R-Factor in the Revised USLE", *Journal of Hydrology*, vol. 157, no. 1, pp. 287–306, 1994.

[REN 97] RENARD K.G., FOSTER G.R., WEESIES G.A. *et al.*, Predicting Soil Erosion by Water: A Guide to Conservation Planning With the Revised Universal Soil Loss Equation (RUSLE), Agriculture Handbook Number 703, United States Department of Agriculture, Agricultural Research Service, 1997.

[ROO 77] ROOSE E.J., "Use of the Universal Soil Loss Equation to Predict Erosion in West Africa", in GREENLAND J., LAL R. (eds), *Soil Erosion: Prediction and Control*, Soil Conservation Society of America, 1977.

[STO 11] STONE R.P., HILBORN D., Universal Soil Loss Equation (USLE) Factsheet Order No. 12-051. Ministry of Agriculture, Food and Rural Affairs, Ontario, 2011.

[TAR 05] TARBOTON D.G., Terrain Analysis Using Digital Elevation Models (TauDEM), Utah State University, Logan, 2005.

[WIS 71] WISCHMEIER W.H., JOHNSON C.B., CROSS B.V., "Soil Erodibility Nomograph for Farmland and Construction Sites", *Journal of Soil and Water Conservation*, vol. 26, pp. 189–193, 1971.

[WIS 78] WISCHMEIER W.H., SMITH D.D., Predicting rainfall erosion losses – a guide to conservation planning, United States Department of Agriculture, 1978.

List of Authors

Rémi ANDREOLI
BLUECHAM SAS
Nouméa
New Caledonia

Nicolas BAGHDADI
Irstea/University of Montpellier
TETIS
Montpellier
France

Flora BRANGER
IRSTEA
HHLY
Lyon
France

Isabelle BRAUD
IRSTEA
HHLY
Lyon
France

Benoit DEFFONTAINES
UPEM-UDD-IGN
LAREG
Marne-la-Vallée
France

Mohammad EL HAJJ
Irstea/University of Montpellier
TETIS
Montpellier
France

Jorge GIRONAS
Pontifical Catholic University
of Chile
DIHA-CEDEUS-CIGIDEN
Santiago
Chile

Tomas GOMEZ
University of Chile
DIC
Santiago
Chile

Nancy HITSCHFELD
University of Chile
DCC
Santiago
Chile

Zeineb KASSOUK
University of Carthage/Institut
National Agronomique de Tunisie
GREEN-TEAM
Tunisia

Michel LE PAGE
IRD
CESBIO
Toulouse
France

Zohra LILI-CHABAANE
University of Carthage/Institut National Agronomique de Tunisie
GREEN-TEAM
Tunisia

Bertrand LUBAC
University of Bordeaux
EPOC-OASU
France

Clément MALLET
IGN-ENSG/University of Paris-Est
MATIS
Saint-Mandé
France

Kenji OSE
Irstea/University of Montpellier
TETIS
Montpellier
France

Paul PASSY
Institut Pierre-Simon Laplace
Paris
France

Hervé PELLA
IRSTEA
UR-Maly
Lyon
France

Fabrice RODRIGUEZ
IFSTTAR/GERS/LEE
IRSTV
Bouguenais
France

Pedro SANZANA
Pontifical Catholic University of Chile
IRSTEA/HHLY
Lyon
France

Adrien SELLES
BRGM
Indo-French Center for Groundwater Research
Hyderabad
India

Valérie SOTI
CIRAD/University of Montpellier
UR AÏDA
Montpellier
France

Ximena VARGAS
Univeristy of Chile
DIC
Santiago
Chile

Sergio VILLAROEL
Univeristy of Chile
DCC
Santiago
Chile

Mehrez ZRIBI
CNRS
CESBIO
Toulouse
France

Index

J

M

N, O

P

Q, R

S

T

V

W, Z

Scientific Committee

Nicolas BAGHDADI, Irstea/University of Montpellier, France

Mehrez ZRIBI, CNRS, France

Clément MALLET, IGN-ENSG/University of Paris-Est, France

Kenji OSE, Irstea/University of Montpellier, France

Mohammad EL HAJJ, Irstea/University of Montpellier, France

Cécile CAZALS, IGN-ENSG/University of Paris-Est, France

Clément DECHESNE, IGN-ENSG/University of Paris-Est, France

Simon BAILLY, IGN-ENSG/University of Paris-Est, France

Marc POUPEE, IGN-ENSG/University of Paris-Est, France

Sébastien GIORDANO, IGN-ENSG/University of Paris-Est, France

Tristan POSTADJIAN, IGN-ENSG/University of Paris-Est, France

Other titles from

iSTE

in

Earth Systems – Environmental Sciences

2018

BAGHDADI Nicolas, MALLET Clément, ZRIBI Mehrez
QGIS in Remote Sensing Set
Volume 1 – QGIS and Generic Tools
Volume 2 – QGIS and Applications in Agriculture and Forest
Volume 3 – QGIS and Applications in Territorial Planning

2017

QUEVAUVILLER Philippe, CIAVOLA Paolo, GARNIER Emmanuel
Management of the Effects of Coastal Storms: Policy, Scientific and Historical Perspectives

2016

BLAIN Stéphane, TAGLIABUE Alessandro
Iron Cycle in Oceans

BRAUDEAU Erik, ASSI Amjad, MOHTAR Rabi
Hydrostructural Pedology

MONACO André, PROUZET Patrick
Seas and Oceans Set
Volume 9 – Tools for Oceanography and Ecosystemic Modeling

QUÉGUINER Bernard
The Biogeochemical Cycle of Silicon in the Ocean

QUEVAUVILLER Philippe
Marine Chemical Monitoring

2015

HERBIN Hervé, DUBUISSON Philippe
Infrared Observation of Earth's Atmosphere

MERCIER-LAURENT Eunika
The Innovation Biosphere

MONACO André, PROUZET Patrick
Seas and Oceans Set
Volume 7 – Marine Ecosystems: Diversity and Functions
Volume 8 – Governance of Seas and Oceans

2014

DE LARMINAT Philippe
Climate Change: Identification and Projections

MONACO André, PROUZET Patrick
Seas and Oceans Set
Volume 1 – Ocean in the Earth System
Volume 2 – The Land-Sea Interactions
Volume 3 – Ecosystem Sustainability and Global Change
Volume 4 – Vulnerability of Coastal Ecosystems and Adaptation
Volume 5 – Development of Marine Resources
Volume 6 – Value and Economy of Marine Resources

MUTTIN Frédéric
Marine Coastal and Water Pollutions: Oil Spill Studies

2013

LÉCUYER Christophe
Water on Earth: Physicochemical and Biological Properties

LEGCHENKO Anatoly
Magnetic Resonance Imaging for Groundwater

2012

LE MENN Marc
Instrumentation and Metrology in Oceanography

PAVÉ Alain
Modeling of Living Systems: From Cell to Ecosystem

2010

TANGUY Jean-Michel
Environmental Hydraulics Series
Volume 1 – Physical Processes and Measurement Devices
Volume 2 – Mathematical Models
Volume 3 – Numerical Methods
Volume 4 – Practical Applications in Engineering
Volume 5 – Modeling Software

Printed and bound by CPI Group (UK) Ltd, Croydon, CR0 4YY
15/09/2026
14962345-0004